Fiber
Optics

second edition

Fiber
Optics

ROBERT J. HOSS
EDWARD A. LACY

 P T R   Prentice Hall, Englewood Cliffs, New Jersey 07632

Hoss, Robert J.
    Fiber optics / Robert J. Hoss, Edward A. Lacy. -- 2nd ed.
        p.   cm.
    Rev. ed. of: Fiber optics / Edward A. Lacy. c1982.
    Includes bibliographical references and index.
    ISBN 0-13-321241-6
    1. Optical communications. 2. Fiber optics. 3. Telecomunication
systems--Design and construction.   I. Lacy, Edward A., 1935-
Fiber optics. II. Title.
TK5103.59.L3 1993
621.382'75--dc20   92-42130
                                                                CIP

Editorial production
  and interior design: *bookworks*
Acquisitions editor: *Karen Gettman*

Cover designer: *Greg Wilkin*
Buyer: *Mary Elizabeth McCartney*

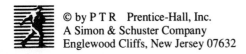

© by P T R   Prentice-Hall, Inc.
A Simon & Schuster Company
Englewood Cliffs, New Jersey 07632

The publisher offers discounts on this book when ordered in bulk quantities. For more information, contact:

Corporate Sales Department
PTR Prentice Hall
113 Sylvan Avenue
Englewood Cliffs, NJ 07632

Phone: 201-592-2863
FAX: 201-592-2249

Printed in the United States of America

10   9   8   7   6   5   4   3   2   1

ISBN 0-13-321241-6

Prentice-Hall International (UK) Limited, *London*
Prentice-Hall of Australia Pty. Limited, *Sydney*
Prentice-Hall Canada, Inc., *Toronto*
Prentice-Hall Hispanoamericana S.A., *Mexico*
Prentice-Hall of India Private Limited, *New Delhi*
Prentice-Hall of Japan, Inc., *Tokyo*
Simon & Schuster Asia Pte, Ltd., *Singapore*
Editora Prentice-Hall do Brasil, Ltda., *Rio De Janeiro*

Contents

Preface

Throughout the world fiber optics is now being used to transmit voice, television, and data signals by lightwaves over flexible hair-thin threads of glass or plastic. In such use it has significant advantages as compared with conventional coaxial cable or twisted wire pairs. Consequently, millions of dollars are being spent to put these lightwave communication systems into operation.

To operate and maintain fiber-optic systems, present-day telecommunications technicians will be forced to learn the basics of fiber-optic components and systems.

This book has been prepared to give the average electronics technician the practical foundation for this challenging innovation. This book is arranged in a logical fashion to describe components and systems. No prior knowledge of optics is necessary.

This text covers the communications use of fiber optics. Other uses—such as medical and industrial inspection of inaccessible areas—are not discussed.

Edward A. Lacy
Satellite Beach, Florida

Since the initial printing of this book, fiber optics has made extensive advances in both component and systems technology. It is now applied in nearly all telephone and data communications networks as the transmission medium of choice. The huge channel capacity of fiber, the long transmission distances possible, and the relatively low cost and simplicity of the electronics has made it the most economical approach for most communications networking applications.

With this in mind I have updated the book to include a revision to some of the latest optical components and their performance characteristics; a full section on the latest systems and network architectures used in the industry; and a new section on installation, maintenance, and measurement techniques useful in the field. Chapter 7 also includes a new section, a primer on systems design. I have tried to make this book applicable to the technician or engineer who has to install and maintain a practical fiber system in today's environment. For this reason such emerging technologies as coherent optics and integrated optics have been considered beyond the scope of this book. I have given them mention, however, as they relate to their effect on the evolution of fiber systems and applications.

Robert J. Hoss
Phoenix, Arizona

Acknowledgments

It is with gratitude that we acknowledge the assistance of the following companies, individuals, organizations, and publishers in the preparation of this book:

AEG-Telefunken
AMP
Amphenol
AT&T Technologies Inc.
Belden Corporation
Bell Canada
Bell Laboratories RECORD
Bell Telephone Laboratories
Capscan
Cablesystems Engineering
Colons Communications
Corning
Dorran Photonics Inc.
Du Pont
EDN (Cahners Publishing Co.)
Electrical Communication
Electronic Design (Hayden Publishing Co., Inc.)
Electro-Optical Systems Design (Milton S. Kiver Publications, Inc.)
EOTec
Fiberguide
Fujitsu

General Cable Company
General Optronics Corporation
GTE Lenkurt
Hamamatsu
Harris Corporation
Hewlett-Packard
Institute of Electrical and Electronics Engineers (IEEE)
International Business Machines (IBM)
International Telephone and Telegraph (ITT)
ITT Cannon
John Wiley & Sons Ltd.
Laser Focus (Advanced Technology Publications)
Machine Design
Math Associates
Mechanical Engineering
Motorola Inc.
National Communication System
Northern Telecom Inc.
Optical Cable Corporation
Optical Information Systems (Exxon Enterprises, Inc.)
Photodyne
PCI
Prentice Hall, Inc.
Professor Morris Grossman
PSI Telecommunications
Quartz & Silice
RCA Corporation
RIFOCS Corp.
Siecor
Siemens AG
TACAN (IPITEK)
Telephone Engineer and Management
Thomas & Betts
3M
Times Wire & Cable Co.
TRW Cinch

Particular thanks to AMP Incorporated for their permission to use numerous drawings, quotes, and paraphrases from their excellent publication, "Introduction to Fiber Optics and AMP Fiber-Optic Products," HB 5444.

1

Fiber Optics: Lightwave Communications

1.1 INTRODUCTION

For years fiber optics has been merely a system for piping light around corners and into inaccessible places so as to allow the hidden to be seen. But now fiber optics has evolved into a system of significantly greater importance and use. Throughout the world it is now being used to transmit voice, television, and data signals by light waves over flexible hair-thin threads of glass or plastic. Its advantages in such use, as compared with conventional coaxial cable or twisted wire pairs, are fantastic. As a result, millions of dollars are being spent to put these lightwave communication systems into operation.

No longer a mere laboratory curiosity, fiber optics is now an important new proven technology, a recognized reality. In fact, some are calling it an exciting revolution that may affect our lives as much as computers and integrated circuits have. In the world of communications fiber optics is compared in importance with microwave and satellite transmission.

It is indeed a new era in communications, the age of optical communications. In many ways it is a radical departure from the electronic communications we have become so accustomed to. Now instead of electrons moving back and forth over metallic wires to carry our signals, lightwaves are being guided by tiny fibers of glass or plastic to accomplish the same purpose.

With a bandwidth or information capacity thousands of times greater than that of copper circuits, fiber optics will provide us with all the communication paths we could ever want, at a price we can afford. It will make practical such services as two-way television that were too costly before the development of fiber optics. In addition to a tremendous bandwidth, fiber optics has smaller and lighter cables than do copper-conductor systems, immunity to electrical noise, and numerous other advantages.

It is no wonder, then, that fiber optics is having a major impact on the electronics industry. Hundreds of companies, both new and old, large and small, are producing fiber systems and components together with government agencies and military services. Thousands of engineers and scientists around the world are now involved in research and development of fiber-optic components and systems. Hundreds of technical papers are being presented as these people continue to make technological breakthroughs.

To date, fiber optics has found its greatest application in the telephone industry. But its other applications for transmitting data are vitally important in many other areas, such as computers, cable television, and industrial instrumentation. Still other uses are expected to be found as the price of fiber-optic systems drops and advanced components are perfected.

Because of the advantages of fiber optics, some designers believe that any new communication system that does not use fiber optics, or at least consider its use, is obsolete even before it has been built. Although this may not always be the case with communication systems, it is becoming more and more obvious that the average technician may also become obsolete if he or she fails to master the basics of this new technology. After all, it will be up to the technician, not the engineer, to repair and maintain fiber-optic systems wherever they are used.

1.2 BASIC THEORY OF OPERATION

Fiber optics can be defined as that branch of optics that deals with communication by transmission of light through ultrapure fibers of glass or plastic. It has become the mainstay or major interest in the world of electro-optics, the blending of the technology of optics and electronics.

In a fiber-optic system or link (Fig. 1.1) three major parts perform this task of communication: a light source, an optical fiber, and a light detector or receiver. The *light source* can be either a light-emitting diode (LED) or a semiconductor laser diode. The *optical fiber* can be a strand as short as 1 m or as long as 10 km. The *detector* can be either an avalanche photodiode (APD) or a positive-intrinsic-negative (PIN) diode. Each of these devices is discussed in detail in later chapters. For now, however, we want to see how they can be combined to form a communication system.

Basically, a fiber-optic system simply converts an electrical signal to an infrared light signal, launches or transmits this light signal onto an optical fiber, and then captures the signal on the other end, where it reconverts it to an electrical signal.

Two types of lightwave modulation are possible: analog and digital. In *analog modulation* the intensity of the light beam from the laser or LED is varied continuously. That is, the light source emits a continuous beam of varying intensity.

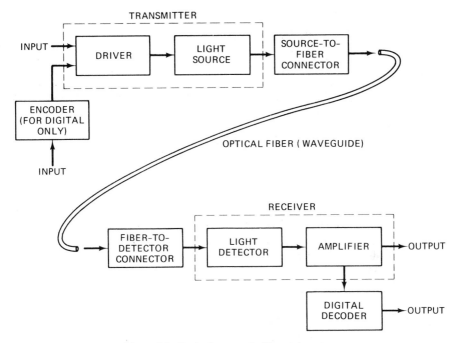

Figure 1.1   Basic elements of a fiber-optic system.

In *digital modulation,* conversely, the intensity is changed impulsively, in an on/off fashion. The light flashes on and off at an extremely fast rate. In the most typical system—pulse-code modulation (PCM)—the analog input signals are sampled for wave height. For voice signals this is usually at a rate of 8000 times a second. Each wave height is then assigned an 8-bit binary number that is transmitted in a series of individual time slots or slices to the light source. In transmitting this binary number, a 1 can be represented as a pulse of light and a 0 by the absence of light in a specific time slice.

Digital modulation is far more popular, as it allows greater transmission distances with the same power than analog modulation. Analog modulation is simpler, however, as shown in Fig. 1.1.

Notice that the encoder and decoder circuits are not necessary for analog modulation. The driver converts the incoming signal, whether digital or analog, into a form that will operate the source.

Even though miniature or tiny light sources and detectors are in use, optical fibers are so small that special connectors must be used to couple the light from the source to the fiber and from the fiber to the detector.

Not shown in Fig. 1.1 are miscellaneous connectors for the optical fiber that allow easier installation and disassembly for repair.

The optical fiber provides a low-loss path for the light to follow from the light source to the light detector. In a sense it is a waveguide that carries optical energy.

Most often this fiber is made of ultrapure glass, although plastic fibers are useful in a few applications. The glass fibers are so pure that they make eyeglass lenses seem opaque in comparison, according to Western Electric engineers. If a window of this material was made 1 km thick, say Bell Canada engineers, it would be as transparent as an ordinary pane of glass.

In many situations, of course, glass is considered to be a hard and brittle substance. However, optical fiber made of glass can be bent (Fig. 1.2) and even knotted, yet it is stronger than stainless steel wire of the same diameter.

An optical fiber used for telecommunications typically consists of a glass core approximately 5 one-thousandths of an inch in diameter. Surrounding this core is a layer of glass or plastic, called a *cladding,* which keeps the lightwaves within the core. Polyurethane jackets are added to the fiber to protect it from abrasion, crushing, chemicals, and the environment.

Individual fibers are often grouped to form cables. A typical fiber-optic cable may contain 1 to 144 of these fibers plus in some cases steel wire fiberglass or Kevlar® that adds strength to the cable. The entire cable is still much smaller in diameter and much lighter than a comparable copper cable.

Although two-way or bidirectional fiber systems are deployed, most fiber systems are strictly one-way per fiber. Thus, two fibers working in pairs are needed for telephone conversations: one strand to transmit a voice from one end of the link, whereas the other carries the voice from the opposite end.

At the end of the fiber, the light signals are coupled to either an APD or a PIN

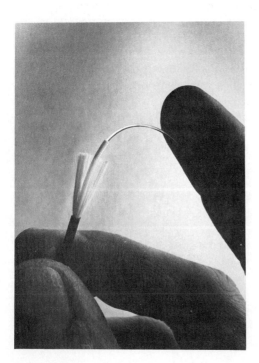

Figure 1.2 Optical fibers can be bent. (Courtesy of DuPont.)

diode. In either device the modulated light signal is converted to an electrical signal which is then amplified and, if necessary, decoded. In decoding, specific time slices or slots are checked for the presence (a binary 1) or absence (binary 0) of a light pulse. These binary digits are combined to form a digital word. In a reverse system to the sampling circuit, digital words are then used to reconstruct the analog wave.

In the system shown in Fig. 1.1, the distance between light source and light detector is short enough that the signal requires no intermediate amplification or regeneration.

When the link becomes too long, the fiber will attenuate the lightwaves traveling down it so that the lightwaves cannot be distinguished from noise. In the early days of fiber optics, this attenuation would occur after only a few meters. Recent advances have extended the range to tens of kilometers before amplification is necessary.

Even with the highest-intensity light sources and the lowest-loss fibers, the lightwaves finally become so weak or dim from absorption and scattering that they must be regenerated. At this point a *repeater* must be placed in the circuit. This device consists of a light receiver, pulse amplifier and regenerator, and a light source. Together they rebuild the pulses to their former level and send them on their way.

Now that we have seen how fiber optics works, let us look at some of its fantastic advantages as compared with coax or twisted-pair communication systems.

1.3 ADVANTAGES

Fiber will be used in a particular application because it exhibits some advantage over an alternative technology. Whenever the characteristics of fiber permit it to be the low-cost alternative, it will generally be employed in an application. Fiber is also employed when it provides a unique feature or service not available with other technologies. If fiber is used in applications where its inherent qualities (such as wide bandwidth and low attenuation) are not required, it will often become a more expensive solution. Some of the primary advantages of fiber over other terrestrial communications mediums are

1. Extremely wide bandwidth
2. Smaller-diameter, lighter-weight cables
3. Lack of crosstalk between parallel fibers
4. Immunity to inductive interference
5. High-quality transmission
6. Low installation and operating costs

But these are merely the *primary* advantages; there are also important *secondary* advantages.

1. Greater security
2. Greater safety

3. Longer life span
4. Environmental stability
5. Electromagnetic pulse (EMP) immunity
6. Rugged construction
7. Greater reliability and ease of maintenance
8. No externally radiated signals
9. Ease of expansion of system capability
10. Use of *common* natural resources

Let us examine first the primary advantages and relate them to the world of telecommunications.

1. *Large length-bandwidth capability.* Fiber today has bandwidth capability theoretically in excess of 10 GHz and attenuations less than 0.3 dB for a kilometer of fiber. This combination of wide bandwidth and low attenuation makes fiber most frequently the lowest-cost transmission medium per channel kilometer. The limits on transmission speed and distance today lies largely with the laser, receiver, and multiplexing electronics.

An extremely wide bandwidth means that a greater volume of information or messages or conversations can be carried over a particular circuit. Whether the information is voice, data, or video, or a combination of these, it can be transmitted easily over fiber optics in large channel groups. Fiber-optic telephone and data transmission systems are in operation today at 4.8 Gb/s. This equates to more than 64,000 simultaneous telephone conversations on a single fiber pair. With the future advent of stable narrow line single-mode lasers and coherent optics, 10 to 100 Gb/s transmission is possible.

2. *Smaller-diameter, lighter-weight cables.* These cables offer obvious advantages with the hair-thin optical fibers. Even when such fibers are covered with protective coatings, they still are much smaller and lighter than equivalent copper cables. For example, a 0.005-in diameter optical fiber in a jacket about 0.25 in in diameter can replace a 3-in bundle of 900 pairs of copper wire [2]. As shown in Fig. 1.3, this enormous size reduction (easily 10:1) allows fiber-optic cables to be threaded into crowded underground conduits or ducts. In some cities these conduits are so crammed they can scarcely accommodate a single additional copper cable.

The size reduction makes fiber-optic cables the ideal transmission system for ships, aircraft, and high-rise buildings, where bulky copper cables take up too much space.

Together with the reduction in size goes an enormous reduction in weight: 208 lb of copper wire can be replaced by 8 lb of optical fiber [3]. It is an important advantage in aircraft, missiles, and satellites.

The combined advantages of smaller-diameter cable and lighter weight give a

Figure 1.3   Size reduction with optical fibers. (Courtesy of Corning.)

decided cost savings in transportation and storage. For instance, the Army has found that a 1¼-ton trailer is sufficient to transport optical fibers of the same capacity as three 2½-ton truckloads of copper cable [2].

Because it is so thin and lightweight, fiber-optic cable can be handled and installed much easier (and with less cost) than copper cable. Thus, there is no need to dig up city streets to lay new conduits.

3. *Negligible crosstalk.* In conventional communication circuits, signals often stray from one circuit to another, resulting in other calls being heard in the background. This crosstalk is negligible with fiber optics even when numerous fibers are cabled together.

4. *Immunity to inductive interference.* As dielectrics, rather than metal, optical fibers do not act as antennas to pick up radio-frequency interference (RFI), electromagnetic interference (EMI), or EMPs. The result is noise-free transmission. That is, fiber-optic cables are immune to interference caused by lightning, nearby electric motors, relays, and dozens of other electrical noise generators that induce problems on copper cables unless shielded and filtered. Carrying light rather than electrical signals, fiber-optic cables ignore these electrical disturbances. Thus, they can operate readily in a noisy electrical environment. They are particularly useful in nuclear environments because of their immunity to EMP effects. Because of their immunity to electromagnetic fields, fiber-optic cables do not require bulky metal shielding and can be run in the same cable trays as power cabling if necessary.

5. *High-quality transmission.* As a result of the noise immunity of the fiber transmission path, fiber routinely provides communications quality orders of magnitudes better than copper or microwave. The general standard for a fiber transmission link is a 10^{-9} BER minimum with 10^{-11} or better as the norm. This is in comparison with 10^{-5} to 10^{-7} for copper or microwave systems.

6. *Low installation and operating costs.* The wide bandwidth and low loss increases repeater spacing, therefore reducing the cost of capital in the outside plant. The elimination or reduction of repeaters reduces maintenance, power, and operating expenses. The simplicity, low power requirements, and very high reliability of the terminal equipment also reduces maintenance costs. Generally fiber can be installed and maintained within the capabilities of an existing telecommunications organization, with little training and only a few added pieces of optical test equipment (a power meter and possibly an OTDR and a splicer). Installation is often simpler than copper systems because fiber requires no special balancing, conditioning, or environmental precautions (pressurization).

In very short applications, it is difficult for fiber optics to compete economically with copper wires. However, where the communication capacity would require coax rather than copper wires, or where interference would require special shielding for metallic wire, fiber links can be competitive even at today's prices [4].

Lifetime costs for a fiber-optic system may be much more attractive and a better basis for comparing fiber optics with wire pairs or coax. Such costs include shipping, handling, and installation as well as manufacture. Before the cable is installed, shipping and handling costs are about one-fourth that of current metal cable, and labor for installation is about one-half less [5].

According to a report by International Resource Development Inc. (IRD), as costs for petroleum and petroleum-related products rise, the use of fiber optics becomes even a more favorable alternative to copper cable because less plastic coating is needed for fiber-optic cable. Still another cost saver, IRD points out, is that connectors used with copper cable are usually gold plated, whereas fiber optic connectors are made from nylon and plastic [6].

Aside from twisted wire pairs and coaxial cable, fiber optics' only other competitor is microwave transmission. Although microwaves can reliably transmit as many bits per second of data, fiber optics promises to be far less costly than microwave towers [1].

By themselves these *primary* advantages are sufficient to justify the use of fiber optics in a number of applications. However, the *secondary* advantages must not be overlooked.

1. *Greater security through almost total immunity to wiretapping.* This is a matter of much greater importance to military services, banks, and computer networks than it is to the average citizen who is calling a relative thousands of miles away. But for these groups, communication security, that is, telephone or data privacy, is well worth any increased cost.

Unless a steel cable is added to a fiber-optic cable for strength, a fiber cable can be laid in an undetectable fashion. It just cannot be found with metal detectors or electromagnetic flux measurement equipment as is the case with wire pairs and coax.

As the light in an optical fiber does not radiate outside the cable, the only way to eavesdrop is to couple light from the fiber physically. If an eavesdropper was smart enough to do this, he or she could force some light (and therefore the message) out, but the loss or disturbance could be detected (with proper electronic sensors) at the receiving end and an alarm sounded.

2. *Greater safety.* Greater safety is available with fiber optics because only light, not electricity, is being conducted. Thus, if a fiber-optic cable is damaged, there is no spark from a short circuit. Consequently, fiber-optic cable can be routed through areas (such as chemical plants and coal mines) with highly volatile gases without fear of causing fire or explosion. In effect, as long as the fiber-optic cable does not have a steel strength member it provides electrical isolation between the transmitter and the receiver. If a cable is disrupted, there can be no short circuits or circuit-loading reflections back to the terminal equipment [1]. In addition, there is no shock hazard with fiber-optic cables. Fibers can be repaired in the field even when the equipment is turned on.

3. *Longer life span.* A longer life span is predicted for fiber optics: 20 to 50 years, compared with 12 to 15 years for conventional cable [7]. Glass, after all, does not corrode as metal does.

4. *Environmental stability.* Fiber retains its transmission characteristics virtually unaffected by environmental extremes encountered in normal installations. Only extreme cold ($-20°C$ to $-40°C$) causes an increase in attenuation. With coax, temperature can have a continuous effect on performance. Water has no effect (except on lifetime if the fiber is under stress), whereas with copper cable it causes rapid performance degradation.

Because fiber-optic cables are made of glass or plastic, in contrast to metal, they have a *high tolerance to liquids and corrosive gases.*

5. *EMP immunity.* Because fiber is not a conductor it provides immunity to EMP. EMP is the result of a high-altitude nuclear burst. The magnetic fields created by the high-speed atomic particles cause free electrons in the upper atmosphere to spin, in turn creating a short duration but high-energy electromagnetic wave. When this wave hits a wire conductor it can produce a short-duration current pulse of 1000 amps or more. The result is the destruction of highly sensitive receiving electronics attached to the wire. To prevent damage, heavy filtering, arc arrestors, and surge protectors are used. Replacing the copper system with fiber not only provides EMP protection, but reduces weight and preserves signal quality. This is particularly important in military aircraft applications and tactical field telecommunications systems.

6. *Rugged construction.* Fiber-optic cable has been designed in many cases to be more rugged than its copper cable counterpart. This is particularly true when comparing it with coax, where the transmission properties can be altered by extreme tension or bending. Fiber cable for outside plant installation generally has a tensile strength of 600 lb. Some tactical field cable designs can withstand multiple tank and vehicle crossings, a property that copper cable cannot provide.

7. *Greater reliability and ease of maintenance.* These are made possible by the extended distance (or spacing) between line amplifiers (repeaters) that boost signal strength. Transmission losses are lower in fiber-optic cable than in coaxial cable, allowing substantial increases between repeaters. Instead of placing a repeater each mile, as in conventional copper wire and coaxial cable systems, repeaters can be positioned 20 miles or more apart. With such distances it is often possible, at least in metropolitan areas, to transmit between telephone exchanges without the use of a single repeater. Obviously, the fewer repeaters there are in a circuit, the less likelihood there will be for circuit failure. The reliability increases accordingly.

In addition to reducing the number of repeaters, the few that are left can be placed indoors. Where the active elements are located within the switching offices, Bell Canada engineers point out, the ease in locating trouble and access to failed equipment is extremely beneficial to improving services and lowering cost.

8. *Signal confinement.* As fiber-optic cables do not radiate signals, fiber-optic transmission does not interfere with other services. Signal confinement is excellent.

9. *System capability.* Many fiber-optic systems can be easily updated to expand system capability. By simply changing terminal equipment and multiplexers, one can upgrade most present-day low-loss fiber-optic systems without replacing the original cable.

10. *Common natural resource.* By using a common natural resource—sand—rather than a scarce resource—copper—fiber optics is helping to conserve a dwindling world resource.

Some of the limitations of fiber are

1. *Radiation hardness.* Fiber darkens to some degree when bombarded by high-energy nuclear particles. If the event is a nuclear blast, the fiber will first glow during the initial event, then darken very rapidly to the point where it loses useful transmission properties. It will then gradually recover, but it will have a higher attenuation than before the event. The recovery level (permanent damage) depends on the level and type of fiber dopant. Lightly doped GeO_2-SiO_2 fused silica fibers such as single-mode or nondoped pure fused silica are affected less. The effect is also much less at longer wavelengths (1300 nm) where loss is very low to begin with. The total transient and permanent absorption is a function of dose and dose rate. For dosage in the 1000 rad range, induced loss of from 20 to 100 dB/km can be expected, and transient recovery can last for less than 1 s to 100 s depending on operating wavelength and fiber doping. Temperature has a strong effect on recovery. Recovery can be 100 times faster for a 50°C increase in temperature, and 100 times slower for a 50°C decrease.

2. *Nonconductor.* Because fiber cannot transmit electrical power, it cannot be used where the receiving terminal (such as a telephone set) must be powered from the line. For such applications the cable must provide separate conductors for power.

3. *Bend strength.* Most fiber is proof tested to between 50,000 and 100,000 lbf/in². The pull tension on fiber therefore is adequate for rough handling when properly cabled and protected. It is difficult, in fact, to pull apart a single-coated fiber with one's bare hands. If a fiber is tightly bent, however, the stress on the outer surface can exceed the tensile strength, and breakage can result. Generally cable is designed to prevent such a small bend radius, and such breakage only occurs with single-fiber "pigtails" or by accident during splice operations.

   4. *Hydrogen absorption.* Molecular hydrogen can diffuse into silica fibers and produce attenuation change. Hydrogen can come from certain cable and fiber buffering materials and, because confined within the cable, can build up to significant concentrations. The losses can come from two mechanisms: (1) unreacted hydrogen, which diffuses into the silica creating reversible losses owing to absorption; and (2) hydrogen molecules reacting with sites in the silica matrix to form permanent hydroxyl ions to a degree dependent on fiber composition, temperature and hydrogen partial pressure [18]. The loss spectrum from the diffusion mechanism has a dominant peak around 1.24 μm with a loss proportional to the hydrogen partial pressure and is reversible if the partial pressure decreases. The most vulnerable to the second mechanism is multimode fiber doped with a high degree of phosphorous. In single mode, any small effect is generally reversible. Under common conditions the effect is generally negligible even in multimode. Charlton quotes some work performed by Rush of BTRL that indicates predicted increases of only 0.05 dB/km at 1300 nm under worst-case cabling conditions of 1000 ppm hydrogen at 20°C for 25 years [18].

   5. *High cost in low bandwidth applications.* Fiber is generally only cost-effective when its capabilities for bandwidth and attenuation are required. If high-quality fiber is used in low bandwidth, short-distance applications, it may cost more than copper. Moderate to high-quality fused silica fiber can cost anywhere from 10 cents to 30 cents per cabled fiber meter, depending on quality, quantity, and fiber count. Lower-quality fiber may cost just as much or more because of the lower production volumes. Copper wire, conversely, is in the $0.03 to $0.30 per conductor meter range (although high-quality coax can exceed this). If the application does not require the dielectric properties of fiber, and runs adequately on copper, then fiber may be a poor choice.

1.4 APPLICATIONS

With all these advantages, there are obviously a lot of places, a wide range of fields, where fiber optics can be used to advantage. This section examines some of the existing applications areas.

1.4.1 Telephone Systems

The world's first optical link providing regular telephone service to the public was placed in operation on April 22, 1977, by General Telephone Company of California. The cable contained 6 fiber pairs, 2 actively carrying voice traffic at a rate of

1.544 Mb/s, which equates to 24 voice channels. It required a repeater every 2 miles and used light-emitting diodes.

Today fiber optics carries telephone transmissions around the globe and regularly operate at 1.2 to 4.8 Gb/s (1.2 to 4.8 thousand Mb/s) over a pair of fibers. Transmission distances are typically 30 to 40 km (18 to 24 miles) between repeaters, and fiber cables typically contain 36 to 144 fibers.

To transmit voice over the fiber network, it is digitally encoded, and the many channels combined using time-division multiplexing. The most common approach is to encode the voice signal with 8-bit-per-sample PCM, resulting in a 64-kb/s digital signal per voice channel. In the United States, Japan, and Canada 24 channels are multiplexed within the PCM channel bank to a 1.544-Mb/s signal known as a T-1. In most of Europe the rate is 2.048 Mb/s for the 24 channels because of two additional 64 kb/s channels used for overhead and synchronization.

Fiber is applied to the telephone network in the interoffice trunking and local distribution sections. There is some fiber being applied to the local loop into individual businesses and even homes, but it will be well into the 1990s until a major penetration is made. Most fiber to customers today is for business customers where bandwidth requirements are large. The specifics of the application to telephone are covered in Chapter 7.

Fiber is the medium of choice for long-distance trunking because it has the most capacity and highest performance for the lowest cost of any other medium. It is also easily upgradeable by changing only the terminal equipment at the ends.

In the local loop, fiber is most often used to interconnect central offices or to provide high-quality services to businesses. Fiber optics is a very popular medium with both the Bell Operating Companies, as well as alternative access carriers, for providing high-speed path-redundant fiber circuits to private business, often dedicating fiber pairs to a single customer. Many new fiber ring tariffs are being developed that provide 565-Mb/s service on route- redundant, automatically recoverable fiber to large businesses [8].

Fiber to the home is still only in the pilot stage, with only a few trial installations in place. The issue here is not a technological one, but one of economics and standards for service delivery. The economics of fiber to the home are more favorable when telephone and Cable Television (CATV) video are combined, but regulatory issues have prevented this for the most part in the United States until just recently. The standards for multiplexing and delivering the signals in a low-cost-per-home fashion must also be developed before much growth will be seen here. Digital encoding of voice and video is the desired approach, but digital encoding is too expensive for the home today. Digital and high-definition television (HDTV) may change the economics in the future, however.

1.4.2 Video

Applications for video transmission on fiber include

1. High-quality video trunked from studio to transmitter
2. Broadcast CATV video

3. Video trunking within a city or between cities
4. Baseband video for closed-circuit, security, or minicam links
5. HDTV trunking and broadcast (fiber to the home)

The most widespread application of video on fiber is in the CATV industry. Warner Amex pioneered the first practical applications of video on single-mode fiber [9] in 1985. Since that time fiber has become the most popular and economical means for trunking multiple channels of video within a metropolitan area. See Chapter 7 (Section 7.5) for further discussion.

1.4.3 Power Stations

Fiber-optic systems are now being used to provide telephone and data communication into and within power stations at the Minnesota Power and Light Company, Florida Power and Light Company, and Georgia Power Company. When copper cables are used for such communication, ground potential rise, induction, and high-frequency arc noise are encountered. But with fiber optics these problems are avoided, as the optical links are simply insensitive to interference [10].

1.4.4 Computers and Data Transmission

The present state of data transmission technology and its evolution has been highly influenced by many years of standardization around early low-bandwidth copper-wire systems. Wideband fiber technology is having a major impact on the way data are transmitted, and although transmission is slowed by these conventional systems standards, totally new standards such as FDDI, SONET, and Asynchronous Transfer Mode (ATM) are emerging, tailored to the broadband capability of fiber.

FDDI is a 100-Mb/s local area network (LAN) standard that permits up to 1000 physical connections in a LAN ring up to 100 km long (see section 7.6.2). SONET is a synchronous time-division multiplexing standard that permits many channels of different formats to be multiplexed together at rates of 2.4 Gb/s or more as well as providing a standard fiber interface (see Section 7.3.3). ATM is a cell relay transmission approach that takes advantage of the wideband transmission capacity that fiber provides.

Fiber plays a role in both public and private data transmission networks as well as private data links between computers and peripherals. In the public network it is a replacement for copper wire and microwave as a transmission medium. The fiber trunks for data transmission are one in the same as the trunks used for voice, described earlier.

When copper-wire or analog transmission systems are used to transmit data, modems are required at either end to convert digital data signals to analog tones that can be transmitted over the analog medium. Fiber is generally implemented as an all-digital transmission medium. In this case digital channel service units and time-division multiplexers are used, which are direct digital-to-digital transmission inter-face and conversion devices. Performance is greater, and costs are lower.

Fiber is also used widely for LANs because of its ability to transmit interfer-ence free, over longer distances. Copper-wire LANs have been limited to the 4- to 10-Mb/s data transmission rate, although rates of 16 Mb/s are now achieved on cop-per. With fiber, LAN speeds of 16 Mb/s and greater are possible. The 100-Mb/s FDDI LAN standard is based on fiber as the medium. See Chapter 7 (Section 7.6.2) for a discussion.

Low-cost LED based fiber links have been used ever since the early 1980s for transmission between processors and peripherals. This is generally done to improve distance and data rate, as well as reduce interference.

In the early 1980s the military was developing fiber for tactical data transmis-sion application. It was applied in aircraft to save weight and reduce problems with EMPs from high-altitude atomic bursts. The Air Force developed some of the early fiber components for military application, pushing into the longer wavelength ranges where attenuation was less and radiation hardness greater [3]. The Navy was the first to deploy fiber optics aboard the A7 aircraft for flight-control testing [11]. Because fiber began to prove itself more rugged than copper, the Army developed the tactical field cable to replace copper twisted pair and coax. The Marine Corps later developed an entire field-deployable fiber-optics databus system (LAN), known as TAOC-85 for tactical air operations. Meanwhile ultra–high-strength fiber was being developed to pay out of the TOW missile, for transmission of video from a camera in the nose of the missile to a controller at the launcher.

1.4.5 Miscellaneous Uses

The following are just a few of the noncommunications uses of active fiber systems:

1. Noncontact temperature measurements
2. Monitoring of current and voltage at high-power stations
3. Barbed-wire perimeter fence that warns of intruders
4. Industrial process control of pit furnaces
5. Measuring hazardous high-level electromagnetic fields without perturbing the field (and thereby distorting the measurements)

1.5 HISTORY

Although the transmission of information by lightwaves over glass or plastic fibers is a relatively new invention, communication by light through the atmosphere is indeed a very old process. In fact, at the end of the sixth century B.C., the news of Troy's downfall was passed by fire signals via a chain of relay stations from Asia Minor to Argos [12].

Centuries later in the United States, American Indians were using smoke signals for communication, and Paul Revere was watching for lantern signals. But by the 1790s, progress was being made in optical communication. Claude Chappe built an optical telegraph system on hilltops throughout France. By means of semaphores, messages reputedly could be transmitted 200 km in just 15 minutes [13]. For its time, it was quite an invention, but by the middle of the nineteenth century, Chappe's telegraph had been replaced by Morse's electric telegraph.

In 1880, Alexander Graham Bell carried his invention of the telephone one step further: instead of transmitting sound waves over wires he used a beam of light. Although Bell considered his photophone to be one of his better inventions, the contraption of mirrors and selenium detectors was cumbersome and unreliable. Also, it had such a limited range that it had no practical value.

Much later, in 1927, Baird in England and Hansell in the United States proposed the use of uncoated fibers to transmit images for television, but their ideas were not pursued [14, p. 1].

It was in the 1930s that single-glass fiber optics were first used for image transmission [15], but the phenomenon was more a laboratory stunt than a practical system.

In the 1950s, studies by A. C. S. van Heel of Holland and H. H. Hopkins and N. S. Kapany of England led to the development of the flexible fiberscope, which is widely used in medical fields [14, p. 1]. Kapany coined the term *fiber optics* in 1956 [14, p. 2].

Optical fibers were developed and placed on the market in the 1950s as light guides that enabled people to peer into otherwise inaccessible places, whether the interior of the human body or the interior of a jet engine. The amount of light lost in these fibers was fantastic, but for the few feet involved in most applications, it did not matter. Still there was no serious consideration given to telecommunication via these fibers because of the high loss of these fibers. Attenuation remained in the 1000 dB/km range even into the mid-1960s.

It was not until 1960 when T. H. Maiman at Hughes Research announced operation of the first laser that optical communications began to look more practical. The laser's high carrier frequency promised a tremendously wide bandwidth for transmitting information.

In theory, a single laser beam could carry several thousand television channels or many many thousands of telephone conversations. Excitement was high. But even though the laser's output could be focused into an extremely narrow and intense beam, it was found that fog and rain could interrupt this beam as it was sent through the atmosphere. Because of this interference, it was actually easier to transmit a reliable laser signal from Arizona to the moon than between downtown and uptown Manhattan [13]. Line-of-sight transmission through the atmosphere, it was concluded, was impractical.

Interest in fiber as the medium began in 1966 when C. Kao and G. A. Hockham at Standard Telecommunications Laboratory predicted that by removing the impurities in the glass, 20 dB/km attenuations would be achievable. At this level fiber becomes a practical communications medium. In 1970 Kapron, Keck, Schultz, and Maurer at Corning broke the 20 dB/km barrier [16]. The fiber-optics revolution had begun.

Suddenly, long-distance telecommunications by fiber optics was possible. More refinements were made to the fibers; cable losses were then cut from 20 dB/km to 6 dB/km and then lower. Room-temperature semiconductor lasers, LEDs, connectors, and photodiodes were developed for use with these low-loss optical fibers. Cable prices began to drop. Within 6 years, working systems were demonstrated and placed on the market.

The first fiber-optics systems in the 1970s were short and used principally for military applications where harsh electromagnetic environments could benefit from the dielectric properties. In 1976, however, Rediffusion of London installed the first commercial system to transmit television signals to its cable TV subscribers. The first prototype telephone systems were also implemented by the Bell System and General Telephone in 1976 and 1977. In 1976 Bell Labs began a major laboratory demonstration of fiber optics for telephones at its Atlanta, Georgia, facility. The system worked over a distance of 10.9 km without repeaters. It used a 144-fiber cable, with each fiber having the ability to handle 672 telephone calls [17].

In one of the first military uses of fiber optics, the U.S. Navy installed a fiber-optic system on the cruiser U.S.S. *Little Rock* for shipboard voice transmission. It was successfully deployed in the Mediterranean for more than 3 years [19].

In the Airborne Light Optical Fiber Technology (ALOFT) program, the navy demonstrated the use of fiber optics in flight tests on board an A-7 test aircraft [11]. Using fiber-optic cables, total cable and connector weight was reduced from 31.9 lb to 2.7 lb [19].

In 1977, the air force successfully demonstrated a fiber-optic link between two tactical command and control communications centers. The dramatic weight and volume reduction made possible by fiber optics in this link showed how military transport requirements could be relieved [3].

In 1977 Corning and NTT reported achieving an attenuation of only 0.5 dB/km at 1200 nm. Systems development based on Graded Index fiber at 3 to 5 dB/km began in 1978 for both military and telephone application. The army and Marine Corps developed rugged fibers and systems for tactical field application. Standards activity began in the military, the SAE, NATO, and the EIA. Although installation of 45-Mb/s and 90-Mb/s systems for telephone-entrance links and interoffice trunks commenced between 1980 and 1982, fiber was not yet cost-effective for full-scale trunking so the market remained small.

The capacity and lower attenuation of single-mode fiber was needed to make long-haul trunking applications practical. In 1983 several factors came together to revolutionize the fiber industry. Deregulation and technology merged and the first volume purchases of single-mode fiber came from MCI at the same time as the first major production of single-mode fiber was occurring at AT&T. The price of single mode was driven almost overnight from about $1.50 per meter to less than $0.50 per meter. The wide-scale commercial fiber revolution had truly begun.

With AT&T, MCI, and other carriers, such as Sprint and the resellers entering the field, the fibering of the United States began. Although there were still military programs, the emphasis on military products faded as the suppliers geared up for the exploding commercial telephone company market. Cabled fiber prices fell to $0.30 per cabled fiber meter, at first held higher by demand but then falling below that figure as the rush began to subside. By 1987 most of the major construction had subsided or leveled off. Fiber had become a commodity, and many companies merged or scaled down operations to stay alive.

By 1985 long-wavelength operation at 1550 nm had become a practical option, although most systems were still designed for 1300 nm. The early 1983 systems operated at 90 Mb/s and 135 Mb/s, with some 405 Mb/s coming from Japan. By 1986 systems in the 560-Mb/s range were being delivered, and by 1988 1.2 Gb/s to 2.4 Gb/s became a possibility. In 1985 R. Hoss of Warner Amex CATV developed the ability to transmit multiple channels of video over a single-mode fiber in a practical CATV environment. Warner installed such systems in Dallas, Texas, and Queens, New York for multiple-channel supertrunking of video to its CATV subscribers (8 channels per fiber). By 1987 companies were advertising up to 16 video channels per fiber. Many fibers to the home experimental systems were implemented in the early to mid-1980s; however, none met the cost per subscriber goals to make it a competitor with coax for the U.S. CATV market. The later part of the 1980s emphasized product standardization, network management systems for fiber networks, and higher-speed multiplexing. All systems until 1989 had operated based on asynchronous multiplexing based principally on DS-1 and DS-3 electrical interface standards; no standard rates and formats existed at the optical interface.

1.6 LOOKING AHEAD

In late 1988 the first-draft SONET standard emerged. This standard will begin the era of synchronous optical communications. It recommended standard synchronous optical and electrical interfaces that drop and insert channels at multiples of the synchronous frame rate. This standard will bring about equipment compatibility at the optical interface as well as open the door for optical multiplexing and switching. In the same year the FDDI 100-Mb/s LAN standard began to emerge, creating a high-speed backbone LAN that could extend throughout a building or into the metropolitan area. With these standards, open-systems architecture at the optical as well as the electrical interface will be possible in the 1990s.

The next major evolutionary phase of fiber will come in the mid- to late 1990s, with the development of coherent transmission. The lasers required for coherent transmission no longer emit a broad spectrum of light; the amount of electromagnetic spectrum that the laser emission occupies is in the megahertz range instead of the terahertz range. Laser light oscillates more like a sinusoidal radio carrier wave at one frequency rather than emitting a band of lightwaves or frequencies as lasers do today. As a result, laser lightwave can be modulated as one would a radio-wave carrier.

With noncoherent optics we treat the optical carrier as an incoherent optical frequency group (not too different from noise) that we envelope modulate, or turn off and on in some fashion that represents signal information. With coherent optical transmission, the sinusoidal optical carrier can be modulated in frequency, amplitude, or perhaps even phase, much like an electrical carrier would be. With radio or electrical transmission we gain the benefit of highly sensitive receivers that are not as affected by thermal noise as are fiber-optic receivers. This is because we can detect electrical or radio-signal modulation with phase and frequency coherent methods that are better at distinguishing signal from noise than are optical detectors alone. In a method known as heterodyning, the receiver has a copy of the transmitted carrier wave that it compares against the received carrier with signal modulation on it. The difference between the two is the signal. When the optical carrier signal is essentially single frequency, heterodyning can be performed in an optical receiver. This results in sensitivity improvements over today's simple envelope detection type of optical receivers. Improvements in sensitivity from 10 to 19 dB over direct envelope detection, by using ASK, FSK, and PSK techniques, have been reported according to Kazovski [17]. This added sensitivity relates to 25 to 50 km further distance without repeaters.

In addition to receiver sensitivity advantages, response transmission capacity can be increased ten-fold or more because the response characteristics of optical modulation components are in the 10-GHz range rather than a few GHz as they are today. Transmission data rates in the 10-Gb/s range become predictable.

This technology will potentially bring about a practical means for carrying 100 channels or more of video, telephone, and data information to the home at a reasonable cost. It will also revolutionize undersea and long-haul trunking.

The reason why this technology will make fiber to the home so inexpensive is that it eliminates the need for all the expensive video and data switching, multiplexing, and encoding electronics required by today's residential prototype systems. With coherent transmission the lasers are virtually single frequency and tunable; thus many optical carriers can be spaced closely together and transmitted on a single fiber. This cannot be done today because the spectral width and controllable spacing of lasers take up too much spectrum, and optical fibers have attenuation and bandwidth operational spectral "windows" that are limited. In most of today's fiber-to-the-home systems only one or a few optical carriers (or colors of light) are used on a fiber entering the home. This means that only a few video or voice signals can enter the home at a time, unless very expensive electrical signal-encoding and multiplexing equipment is used to put more signals on each optical carrier. Today one must spend a lot of money on either expensive multiplexing and encoding equipment or expensive switching equipment at nodes within the network to select signals that each home desires at any one time.

With coherent optics it is possible for 100 or more optical carriers to appear on the fiber entering the home at one time. With each carrier representing a single television or telephone or data signal, then external switching or multiplexing equipment is not needed. One needs only to "tune" to the optical carrier of choice and demodulate the signal desired (much as a CATV converter does today). This would be accomplished with a relatively low complexity optical tuner and receiver. The tuner would work like a heterodyne radio receiver in that it would contain a "tunable" narrow spectrum laser that would be electrically or mechanically tuned to nearly match the frequency of the chosen optical carrier. The two signals would be mixed, and the resulting "difference frequency" waveform along with the information signal would be processed and the information extracted by conventional electronics similar to that in a standard AM or FM radio. The method is very simple and potentially inexpensive.

Coherent optics requires several advances in component production capability. Laser spectral line widths in the 10-kHz to 1-MHz range are required. This has been demonstrated, however, in external cavity-type lasers and to a lesser degree with distributed feedback lasers. Low-loss external modulators or narrow line lasers that can be directly modulated are required. Mixing couplers and polarization control is also required. External modulators of $LiNbO_3$ have been around for a long time; however, they have been lossy. Newer materials are evolving, which should reduce this problem. Laser stability and frequency control are issues to be resolved. The performance advantages of coherent optics as well as their flexibility will make the development worthwhile and will provide the next great evolutionary step in fiber optics.

1.7 OPPORTUNITIES FOR TECHNICIANS

To be a part of this new technology, to operate, test, and maintain these fiber-optic systems, you need to learn the basics of fiber optic components and systems.

The remainder of this book has been prepared to give you the necessary foundation for this challenging innovation. Starting with a discussion of the fundamentals of light, we proceed in a logical fashion from transmitter to receiver to describe components and systems.

REFERENCES

1. Fritz Hirschfeld, "The Expanding Industry Horizons for Fiber Optics," *Mechanical Engineering,* Jan. 1978, pp. 20–26.
2. "Introduction to Fiber Optics and AMP Fiber-Optic Products," HB 5444, AMP Incorporation, Harrisburg, Pa., n.d.
3. C. W. Kleekamp and B. D. Metcalf, "Fiber Optics for Tactical Communication," Mitre Corp., Bedford, Mass., Apr. 1979, p. 13.
4. Ira Jacobs, "Lightwave Communications—Yesterday, Today, and Tomorrow," *Bell Laboratories RECORD,* Jan. 1980, p. 7.

5. Glen R. Elion and Herbert A. Elion, *Fiber Optics in Communications Systems* (New York: Marcel Dekker, Inc., 1978).

6. "Gold, Copper Costs to Spur Fiber Optic Connector Market?" *Electro-Optical Systems Design,* Apr. 1980, p. 16.

7. News Release, Harris Corporation, Melbourne, Fla., Jan. 16, 1980 (?).

8. Bell South, "Smart Ring" tariff filed Nov. 1991. US West 565-Mb/s fiber access installed in Phoenix in 1989 for a private customer.

9. F. McDevitt and R. Hoss, "Repeaterless 16 km Fiber Optic CATV Supertrunk Using FDM/WDM," paper presented at the NCTA 32nd Annual Convention and Exposition, Houston, June 12–15, 1983.

10. C. A. Ebhardt, "FTS-1, T1 Rate Fiber Optic Transmission System for Electric Power Station Telecommunications," paper presented at the Second International Fiber Optics and Communications Exposition, Chicago, Sept. 5–7, 1979.

11. John D. Anderson and Edward J. Miskovic, "Fiber Optic Data Bus," Northrop Corp., Hawthorne, CA and ITT Cannon, Santa Ana, CA, n.d.

12. P. Russer, "Introduction to Optical Communications," in *Optical Fibre Communications,* ed. M. J. Howes and D. V. Morgan (Chichester, England: John Wiley & Sons Ltd., 1980), p. 1.

13. W. S. Boyle, "Light-Wave Communications," *Scientific American,* Aug. 1977, p. 40.

14. N. S. Kapany, *Fiber Optics, Principles and Applications* (New York: Academic Press, Inc., 1967), p. 1.

15. *The Optical Industry and Systems Directory, Encyclopedia Dictionary,* 24th ed., The Optical Publishing Co., Inc., Pittsfield, Mass, 1978.

16. Steward E. Miller, Enrique A. J. Marcatili, and Tingye Li, "Research toward Optical Fiber Transmission Systems," *Proceedings of the IEEE,* Vol. 61, No. 12, Dec. 1973, pp. 1703–1751; from p. 1704.

17. Kenneth J. Fenton, "Fiber Optics: Claims Are Being Substantiated," *Information Display,* Jan. 1979.

18. Discussions with David Charlton of Corning.

19. "Fiber Optics—Expanding Fleet Capability," Naval Ocean Systems Center Technical Document 224, Mar. 1979.

2

Fundamentals
of Light

As many electronics technicians have had no education or experience in optics, in this chapter we discuss those fundamental characteristics and phenomena of light that relate to fiber optics. We limit the discussion to those principles of light that are relevant to a basic understanding of optical fibers, LEDs, injection lasers, and photodetectors. These principles are necessary for understanding fiber-optic components discussed later. Detailed mathematical explanations of lenses, quantum mechanics, and the like will not be given, as they are needed more by the physicist than by the technician.

2.1 NATURE OF LIGHT

2.1.1 Wave vs. Particle Theory

For centuries, physicists have attempted to describe the nature of light to provide us with a simple picture or model, just as the model of the atom helps us to understand electronics. But there is no simple description, no single theory that explains it completely. However contradictory it may sound, light has the characteristics of particles and of waves.

At times, light seems to be a stream or rain of fast-moving electromagnetic particles called *photons*. Although they are called particles, they bear little resemblance to material particles. For example, they move at fantastic speeds, but at rest they have zero mass. That is, a photon at rest is nonexistent. Instead of calling them particles, perhaps it would be better to call them discrete bundles or packets of energy.

Using the *particle theory*, physicists can describe what happens to light when it is emitted or absorbed. In particular, this theory explains the photoelectric effect, when light striking the surface of certain solids causes them to emit electrons. Without this theory, the behavior of light during emission and absorption cannot be adequately described. But it still does not explain many other phenomena of light.

For example, in numerous experiments light appears to behave like electromagnetic waves instead of a stream of photons. These electromagnetic waves consist of oscillating electric and magnetic fields. Each field is at right angles to each other and to the direction of travel or propagation.

The strength of each field varies sinusoidally. Because these fields oscillate at right angles to the direction of propagation, lightwaves are said to be *transverse*. Like other electromagnetic waves, light can travel through empty space and over very great distances.

The *wave theory* best explains light propagation or transmission. It also explains why light beams can pass through one another without disturbing the other. Notice what happens when two searchlight beams cross each other. Each proceeds from the intersection as if the other had not been there [1]. With particles, this phenomenon could not occur.

Consider, too, the phenomenon of interference. If light from a single source is split into two beams and if the two beams travel over two unequal paths to reach a common point, the beams will interfere with each other. Depending on their phases, these beams will either increase or decrease the intensity of the light at the common point. This characteristic can only be explained by the wave theory.

Therefore, to describe the nature of light completely we must use both the particle theory and the wave theory. Most often we will use the wave theory, but remember that neither theory is sufficient in itself. Both theories can be used, depending on the problem. But to make a perfect mental picture or model of light is just not possible at this time.

2.1.2 The Wavelength of Light

If light is an electromagnetic wave, what then is its frequency or wavelength?

In most electronics work it is customary to refer to electromagnetic waves in terms of *frequency* rather than *wavelength*. In optics, however, wavelength is the important term, as it can be measured directly.

The wavelength of light waves is most typically measured in nanometers (nm), micrometers (μm), and angstroms (Å), instead of feet or inches. (In some literature, meter may also be spelled "metre.") The relationship of these units is as follows:

Unit	Old Designation	Dimensions	
micrometer[a]	micron	10^{-6} m or	10^{-4} cm
nanometer[a]	millimicron	10^{-9} m	10^{-7} cm
angstrom		10^{-10} m	10^{-8} cm

[a] 1000 nm = 1 μm.

The old designations, unfortunately, are still being used in some books and articles, even though they are obsolete. All three units are commonly used by optics workers, but *nanometers* is the unit used most commonly in fiber optics. A typical wavelength encountered in fiber optics is 820 nm; this, of course, may be expressed also as 0.820 μm or 8200 Å.

To give you an idea of the extremely small wavelengths involved in fiber optics, consider the following scale:

$\vdash\!+\!\dashv$ 20 mm

The distance between each mark is 1 mm. A wavelength of 820 nm therefore would be 820 millionths of this distance, or 0.0000323 in! Fig. 2.1 shows the scale of lightwave communications.

2.1.3 Electromagnetic Spectrum

Fundamentally, there is no difference between lightwaves and other electromagnetic waves, such as radio and radar, except that lightwaves are much shorter and therefore have a much higher frequency. When all types of electromagnetic radiation are ar-

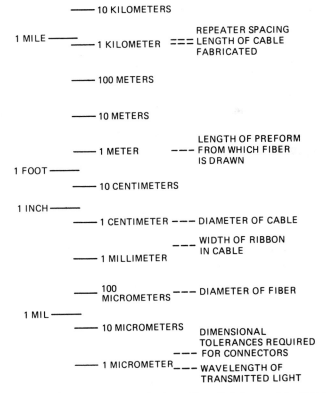

Figure 2.1 Scale of lightwave communications. (From Ref. 2; copyright 1976 Bell Laboratories RECORD.)

ranged in order of wavelength, the result is called the *electromagnetic spectrum,* shown in Fig. 2.2.

Notice how broad this range is: from long electrical oscillations with wavelengths measuring thousands of kilometers to cosmic rays with wavelengths in trillionths of a meter. There are no gaps in the spectrum, but some of the regions overlap or blend. That is, the boundary between regions is not sharp.

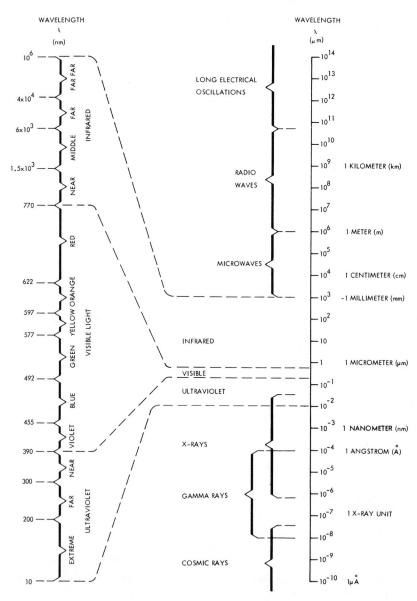

Figure 2.2 Electromagnetic spectrum. (From Ref. 3; copyright 1974 RCA Corporation.)

As can be seen in Fig. 2.2, optical radiation lies between microwaves and x-rays. It includes all wavelengths between 10 nm and 1 mm. Within this range, as shown on the left side of Fig. 2.2, are ultraviolet, visible light, and infrared radiation. The term *visible light* seems redundant; however, it is necessary because ultraviolet and infrared radiation are referred to as ultraviolet light and infrared light, respectively, in some textbooks. *Visible* light is defined as that radiation which stimulates the sense of sight (that is, affects our optic nerves). It includes all radiation from 390 to 770 nm, from violet to red. Light itself does not have color, but these wavelengths stimulate color receptors in the eye. Obviously, the visible spectrum is just a small fraction of the electromagnetic spectrum.

In fiber optics, a typical wavelength is 820 nm. From Fig. 2.2 we see that this radiation is designated infrared, although it is sometimes referred to as light because it can be controlled and measured with instruments similar to those used for visible light.

2.1.4 The Speed of Light

Like other electromagnetic waves, light travels through free space (that is, a vacuum) at the fantastic speed of 186,000 miles per second or 300,000,000 (3×10^8) meters per second. For precision work, the figure of 2.997925×10^8 meters per second (m/s) should be substituted for this figure. But for most practical or ordinary uses, the round number of 3×10^8 m/s is satisfactory. Light traveling in the atmosphere moves slightly slower than this, but the figure of 3×10^8 m/s is still sufficiently accurate.

For propagation in free space and in the atmosphere, the speed of light is the same for all wavelengths; however, in other materials, such as water and glass, different wavelengths travel at different speeds. Regardless of the wavelength, when light travels through such materials, its speed is noticeably reduced. Because of this slowing down, a light ray moving from air into a solid or a liquid will bend at the surface of the new medium. (In optics, a medium is any substance that transmits light.) Such bending or refraction is very important to the study of fiber optics, so it will be discussed in more detail in a later paragraph.

The speed of light has an importance far beyond the field of optics: It is one of the fundamental constants of nature. As such, it crops up in numerous scientific equations, some of which have nothing to do with light. In equations it is designated as the constant c.

2.1.5 Straight-Line Propagation

For most practical purposes, such as carpentry and navigation, light can be considered to travel in an essentially straight line as long as it stays in a uniform medium. (A uniform medium is simply a substance that has uniform composition throughout.)

Any bending of a light ray traveling through the atmosphere is so slight that it can be ignored in most measurements that are based on light. Thus, light is said to have the property of *rectilinear propagation* when it is traveling through a uniform medium.

When it travels from one medium to another, however, it will change its direction or bend at the boundary between the two mediums. Just how much it will bend is discussed later in the section "Refraction." But once it enters a second uniform medium, it will continue in a straight line as long as it stays in that medium.

2.1.6 Transparent, Translucent, and Opaque Materials

Liquids, gases, and solids can be classified as transparent, translucent, or opaque, depending on how much light can penetrate or pass through them. If light can pass through a material with little or no noticeable effect, the material is called *transparent*. In this category we would place water, air, some plastics, and glass. The glass used for optical fibers is so ultrapure, according to Bell Laboratory scientists, that if seawater were so clear, you could see to the bottom of the deepest ocean.

If no light can pass through a material, the material is said to be *opaque*. But even transparent materials can become opaque if you increase the thickness or number of layers sufficiently. Notice that clear water at the edge of a lake where the water is shallow is transparent for a few feet from the shore. In the center of the lake, however, at depths of hundreds of feet, no light can pass through.

Some items—wax paper, for example—will pass light but you cannot see clearly through them. These items are called *translucent*.

2.1.7 Rays of Light

In designing and analyzing optical equipment, it is often desirable to draw simple diagrams showing lenses and other optical devices. But how should the path of the light through the equipment be drawn?

Consider a small light source that is radiating light waves in all directions. These waves can be pictured as spherical surfaces concentric with the source. As the distance of the wave from the source increases, the curve begins to flatten out, forming a straight line. Thus, when the distance becomes large enough, the spherical surfaces can be considered planes. The result is a train of plane waves [4].

A train of wavefronts can be drawn as shown in Fig. 2.3. The drawing can be further simplified by replacing the train of light waves with a single straight line called a *ray* (Fig. 2.4), which is drawn in the direction in which the waves are traveling.

Optical problems can often be solved using such rays and determining their angles and relationships through the use of geometry. This approach is called *geometrical optics*. As it is the simplest approach, it is widely used in the design of optical equipment.

In contrast, *wave optics* considers only the wave properties of light in the treatment of optical problems.

Figure 2.3 Train of wavefronts.

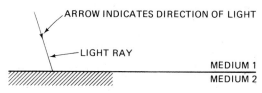

Figure 2.4   Light ray.

2.2 REFLECTION

When a light beam strikes an object, some or all of the light will bounce off or be turned back; the light is said to be *reflected*. If the surface of the object is smooth and polished, as with a sheet of silver, *regular* reflection or *specular* reflection will occur, as shown in Fig. 2.5.

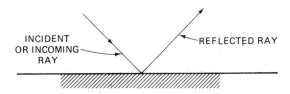

Figure 2.5   Regular or specular reflection.

If the surface of the object has irregularities or is rough in comparison to the wavelength of light, as is usually the case, the light will be reflected in many directions, as illustrated in Fig. 2.6. This case represents *diffuse* reflection.

In everyday use, the ordinary mirror illustrates specular reflection, whereas most nonluminous bodies demonstrate diffuse reflection. Without diffuse reflection we would be unable to see such bodies.

Let us now add an imaginary line called the *normal* to Fig. 2.5 and obtain Fig. 2.7. Note that the normal is perpendicular to the surface. The angle *i*, formed by the normal and the incident ray, is called the angle of incidence. The angle *r* is called the angle of reflection. By the *law of reflection,* the angle of incidence equals the angle

Figure 2.6   Diffuse reflection. (Courtesy of George Shortley and Dudley Williams, *Elements of Physics,* 3rd ed., p. 512; copyright 1961; reprinted by permission of Prentice Hall, Inc., Englewood Cliffs, N.J.)

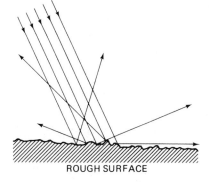

ROUGH SURFACE

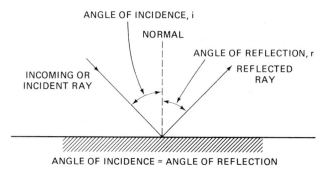

Figure 2.7 Law of reflection.

of reflection. That is, $i = r$. Furthermore, the incident ray, the reflected ray, and the normal are always in the same plane.

2.3 REFRACTION

When light strikes a surface head on, as shown in Fig. 2.8, part of it will be reflected (not shown) and part will be absorbed, as shown by the penetrating ray. We assume in this case that the material will transmit light. For this discussion, we ignore the reflected ray and concentrate on the penetrating ray. As long as the incident ray is perpendicular to the surface, it will continue in a straight line in the new medium, as shown in the figure.

Even though the penetrating ray will not change directions in this particular case, it will slow up appreciably. In free space or air, as you will recall, the velocity of light is 186,000 miles/s. However, in most substances it is less. For instance, in water it is 140,000 miles/s and in an optical fiber it is 124,000 miles/s.

Now consider the case when the light ray is not head on but at an oblique angle, as illustrated in Fig. 2.9. Instead of incident ray *AB* continuing in a straight line as *BC*, it changes its direction to *BD*. (If the angle of incidence is greater than the critical angle, which will be described later, refraction will not occur.) This bending or refraction is caused by the change of speed of the ray as it enters medium 2. In this

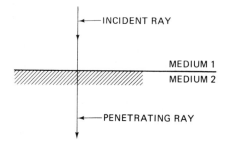

Figure 2.8 Light ray penetrating a second medium.

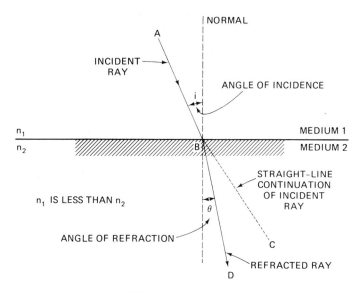

Figure 2.9   Refraction.

particular case, medium 2 is more dense than medium 1, and therefore the refracted ray bends toward the normal. (If medium 1 had been more dense than medium 2, the refracted ray would bend *away* from the normal, and line *BD* would lie on the other side of *BC*.)

The angle *i*, formed by the incident ray and the normal, is the angle of incidence. For refraction to take place, this angle must be greater than 0 degrees and less than 90 degrees. The angle θ, formed by the normal and the refracted ray, is the angle of refraction.

Before we look at the relationship of the angle of incidence to the angle of refraction, we need to look at the *index of refraction* (also called the refractive index), which is designated by the letter *n*. This index is simply the ratio of the speed of light in air (*c*) to the speed of the light being considered (*v*), or

$$n = c/v$$

Typical indexes of refraction are given in Table 2.1. As we have noted earlier, different wavelengths of light travel at different speeds. The typical indexes are based on a wavelength of 5890 Å, the wavelength of a sodium flame.

When a light ray traveling in a medium with an index of refraction of n_1 strikes a second medium (with an index of refraction of n_2) at an angle of incidence *i*, the angle of refraction θ can be determined very easily by Snell's law:

$$n_1 \sin i = n_2 \sin \theta \tag{2.1}$$

Notice that if medium 1 is air, $n_1 = 1$ and can be dropped from the equation, leaving

$$\sin i = n_2 \sin \theta$$

Just as in the case with reflected rays, the incident ray, the normal to the surface at the point of incidence and the refracted ray will all be in the same plane.

TABLE 2.1 TYPICAL INDEXES OF REFRACTION

Air	1.00
Diamond	2.42
Ethyl alcohol	1.36
Fused quartz	1.46
Glass	1.5–1.9
Optical fiber	1.5
Water	1.33

In our diagrams up to this point, we have shown the incident ray and the refracted ray going down on the page. However, this path can be exactly reversed and the ray can go up instead of down. In such a case equation (2.1) still applies as long as the incident ray is defined as the ray in the less dense medium and the refracted ray as the one in the more dense medium.

When a ray of light is moving from a dense medium (high index of refraction) to a less dense medium (lower index of refraction) it will *not* be refracted if it strikes the surface at an angle equal to or greater than a particular angle called the *critical angle*. Instead, it will be totally reflected at the surface between the two media.

To see how this phenomenon of total internal reflection occurs, recall that a light ray will bend *away* from the normal if it is moving into a less dense medium, as shown by ray 1 in Fig. 2.10. As the angle of incidence is increased, from i_1 to i_2, notice that the angle of refraction increases so that ray 2 is closer to the boundary between the two media. (Although not shown in Fig. 2.10, reflection also occurs until the critical angle is reached.)

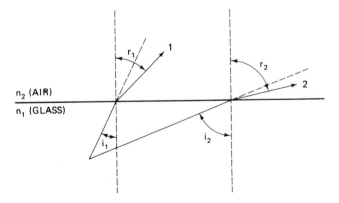

Figure 2.10   Light ray bending away from normal.

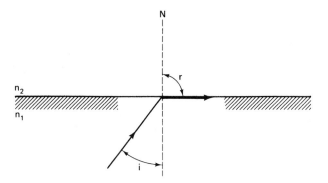

Figure 2.11   Total internal reflection: $i=\theta_c$=critical angle.

When the angle of incidence is increased to a high enough value, called the critical angle (θ_c), the refracted ray just grazes the surface and travels parallel to it, as shown in Fig. 2.11. At this point, the angle of refraction r is 90 degrees.

Substituting in Snell's law, we obtain

$$n_1 \sin r = n_2 \sin 90 \text{ degrees}$$

$$= n_2(1)$$

$$\sin r = \frac{n_2}{n_1}$$

But

$$\sin r = \sin \theta_c$$

Therefore,

$$\sin \theta_c = \frac{n_2}{n_1}$$

Thus, for any ray whose angle of incidence is greater than this critical angle, total internal reflection will occur at the surface between two media provided that the light ray is traveling from a medium of higher refractive index to a medium with a lower refractive index.

Dispersion.   Up until this point, we have assumed, although we may not have stated it, that our light beam or ray consisted of only one wavelength. Such light, which is called _monochromatic,_ is not naturally encountered in the real world. Most light beams are complex waves which contain a mixture of wavelengths and thus are called _polychromatic._ As shown in Fig. 2.12, white light can be separated into individual wavelengths by a glass prism through the process of _dispersion._

Notice that white light is actually a combination of six colors. Dispersion is based on the fact that different wavelengths of light travel at different speeds in the

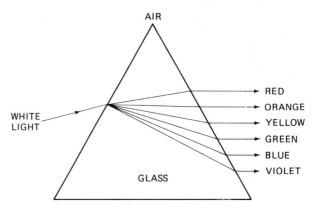

Figure 2.12 Dispersion.

same medium. Remember that in Table 2.1 we specified a particular wavelength. Because different wavelengths have different indexes of refraction, some will be refracted more than others.

  Propagation in Optical Fibers. An optical fiber is produced by forming concentric layers of cladding glass around a core region. The core region maintains the low optical loss properties necessary for the propagation of the optical energy. As we learned previously, the higher the refractive index, the slower optical energy will propagate. In manufacturing an optical fiber a higher refractive index glass core is surrounded by a lower refractive index–cladding material. The result is an optical waveguide whereby the light energy is contained within the higher index core by the internal reflection setup at the core-to-clad interface, as discussed earlier in this section. Further discussion is included in Chapter 4.

2.4 LIGHT MEASUREMENTS

Two systems of optical measurements are encountered in fiber optics: radiometry and photometry. *Radiometry* applies to the measurement of radiant energy of the entire spectrum, regardless of wavelength. *Photometry* is concerned only with measuring that part of the spectrum that affects the eye. That is, photometry does not consider radiation that the eye cannot see.

  Although these systems use different terms and symbols, they are related and conversion can be made from one system to the other, as shown in Tables 2.2 and 2.3.

  Numerous terms are *not* included in these tables because there has been little standardization between the various technical societies, industry, and publishers. The distinction between some of the terms is often quite subtle and confusing. And some of the so-called obsolete terms refuse to go away. But by going back to the basic units—lumens and watts—you can resolve many of the conversion problems.

TABLE 2.2 RADIOMETRIC AND PHOTOMETRIC UNITS

| Definition | Radiometric | | Photometric | |
	Name	Unit (SI[a])	Name	Unit (SI[a])
Energy	Radiant energy	joule	Luminous energy	lumen-second
Energy per unit time = power = flux	Radiant flux	watt	Luminous flux	lumen
Power input per unit area	Irradiance	W/m^2	Illuminance	lm/m^2 lux
Power per unit area	Radiant exitance	W/m^2	Luminous exitance	lm/m^2
Power per unit solid angle	Radiant intensity	W/steradian	Luminous intensity	candela
Power per unit solid angle per unit projected	Radiance	W/m^2-steradian	Luminance	$candela/m^2$

[a]International System of metric units—recommended standard.

Source: Ref. 5; courtesy of M. Grossman.

TABLE 2.3 CONVERSION TABLES FOR ILLUMINATION AND LUMINANCE

Illumination conversion factors[a]

1 lumen - 1/680 lightwatt (at 555 nm)
1 lumen-hour = 60 lumen-minutes
1 footcandle = 1 lumen/ft^2

1 watt-second = 1 joule = 10 ergs
1 phot = 1 lumen/cm^2
1 lux = 1 lumen/m^2

Number of → Multiplied by → Equals number of	Footcandles	Lux	Phots	Milliphots
Footcandles	1	0.0929	929	0.929
Lux	10.76	1	10,000	10
Phots	0.00108	0.0001	1	0.001
Milliphots	1.076	0.1	1,000	1

Luminance conversion factors

1 nit = 1 candela/m^2
1 stilb = 1 candela/cm^2
1 apostilb (international = 0.1 millilambert = 1 blondel
1 lambert = 1,000 millilamberts

Number of → Multiplied by → Equals number of	Foot-lamberts	Candelas /m^2	Milli-lamberts	Candelas /in^2	Candelas /ft^2	Stilbs
Footlamberts	1	0.2919	0.929	452	3.142	2.919
Candelas/m^2	3.426	1	3.183	1,550	10.76	10,000
Millilamberts	1.076	0.3142	1	487	3.382	3,142
Candelas/in^2	0.00221	0.000645	0.00205	1	0.00694	6.45
Candelas/ft^2	0.3183	0.0929	0.2957	144	1	929
Stilbs	0.00034	0.0001	0.00032	0.155	0.00108	1

[a]Footcandle is an obsolete term.

Source: Ref. 5; courtesy of M. Grossman.

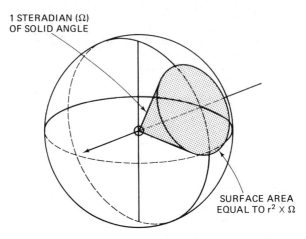

Figure 2.13   Steradian. (From Ref. 5; courtesy of M. Grossman.)

The *watt* is the fundamental unit in radiometry, and the *lumen* is the fundamental unit in photometry. At the eye's peak sensitivity, at a wavelength of 555 nm, 1 watt = 680 lumens.

The 16th General Conference on Weights and Measures has defined the *candela* as the luminous intensity, in a given direction, of a source emitting monochromatic radiation of frequency 540×10^{12} Hz and whose radiant intensity in this direction is 1/683 W/sr [6].

A steradian is defined in Fig. 2.13. Fig. 2.14 shows some of the relationships of the terms in Table 2.2. Fig. 2.15 shows light-level brightness values of common objects.

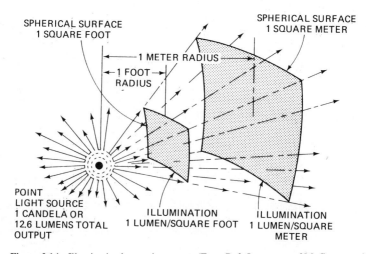

Figure 2.14   Illumination by a point source. (From Ref. 5; courtesy of M. Grossman.)

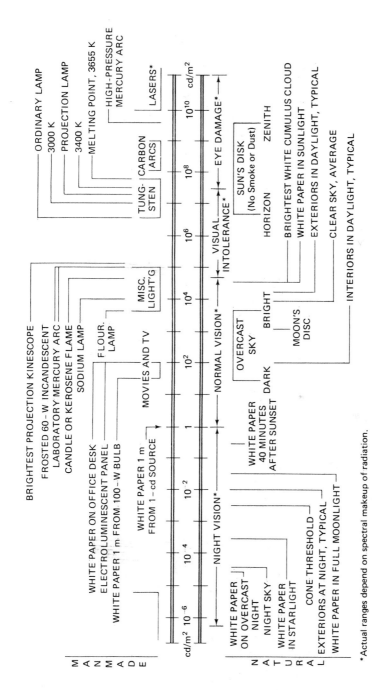

Figure 2.15 Luminance ranges. (Reprinted with permission from Ref. 7 [*Electronic Design*, Vol. 26, No. 4]; copyright Hayden Publishing Co., Inc., 1978.)

*Actual ranges depend on spectral makeup of radiation.

REFERENCES

1. Physical Science Study Committee, *Physics* (Lexington, Mass.: D. C. Heath and Company, 1960), p. 257.
2. Ira Jacobs, "Lightwave Communications Passes Its First Test," *Bell Laboratories RECORD,* Dec. 1976, p. 297.
3. *RCA Electro-Optics Handbook,* p. 14.
4. Francis Weston Sears, *Optics* (Reading, Mass.: Addison-Wesley, 1949), p. 4.
5. Morris Grossman, *Technician's Guide to Solid-State Electronics* (Englewood Cliffs, N.J.: Prentice Hall, Inc., 1976).
6. "Photometric Units Redefined," *Electro-Optical Systems Design,* Apr. 1980, p. 16.
7. "Focus on Lamps," *Electronic Design,* Vol. 26, No. 4, Feb. 15, 1978, p. 58.

3

Light Sources and Transmitters

Numerous devices are available for converting electronic signals to light waves for fiber-optic telecommunication systems. However, at present only two of these devices are really suitable for fiber optics: the LED and the injection laser diode (ILD). Both are semiconductor diodes that are directly modulated by varying the input current. They are close relatives, both being made of aluminum-gallium-arsenide (GaAlAs), gallium-aluminum-arsenide-phosphide (GaAlAsP), or gallium-indium-arsenide-phosphide (GaInAsP).

3.1 OPTICAL SOURCES

The LEDs used in fiber optics are similar to those LEDs used in numerous display applications with the exception that the display devices emit in the visual region (600- to 800-nm wavelength), and those used for communications emit in the infrared (800- to 1300-nm wavelength). ILDs are similar to those used in compact disk players and bar-code readers, except that the emission spectrum is narrower, and operation is in the infrared.

The ILD is also referred to as a laser diode or diode laser.

LEDs are typically used for shorter, slower systems such as LANs and computer peripheral data links. ILDs are used for the faster, longer-length systems. This is because LEDs have larger active surface areas and wider beam angles that make them better able to couple into larger-core multimode fibers. They are relatively moderate to slow speed (typically less than 100 Mb/s. They also emit light over a broad spectrum of wavelengths (30 to 60 nm wide), which creates a bandwidth problem over long distances because of a phenomenon called *material dispersion* (see Chapter 4). ILDs conversely, have tiny active regions and small beam angles that couple well with the lowest loss single-mode fibers. Their higher speeds (greater than 1 Gb/s) and narrow spectral emission are ideal for the wide bandwidth demands of long-distance single-mode fiber communications.

Table 3.1 illustrates some of the common performance parameters for LEDs and ILDs [1].

Both devices have unique advantages. The final choice between them in any given application depends on cost, optical power levels, modulation rates, wavelength, temperature sensitivity, coupling efficiency, and lifetime. These factors will be considered briefly before we discuss each device in detail.

3.1.1 Source to Fiber Coupling

LEDs and ILDs for fiber optics are very small devices, the chips being on the order of a grain of sugar in size. Their active regions closely match the dimensions of the fiber cores that they couple to. Fig. 3.1 illustrates the actual size of packaged devices without coupled fibers.

Ideally, for high efficiency all light from the light source should enter the fiber. The emitting region, however, is often larger than the core area of the fiber and, as illustrated in Fig. 3.2, the light rays are not parallel. As will be discussed in Chapter 4, fiber can only accept light that falls within a narrow acceptance cone, which is 30 to 40 degrees for multimode fiber and less than 10 degrees for single mode. LEDs typically emit light rays in a 120- to 180-degree pattern, which means that most of the light is not accepted by the fiber even if the core and emission area match. ILDs, conversely, emit in a much narrower pattern (typically 10 to 35 degrees) and therefore can couple more of their light into a fiber.

To achieve optimum coupling, sources generally come with fiber "pigtails" attached as illustrated in Fig. 3.3. In the case of LEDs that emit from their top surface, the surface is etched so that the fiber can be placed close to the emission region. In the case of the ILD (and edge-emitting LEDs), which emits from its edge, the fiber is often rounded at the end to form a lens that improves coupling.

ACTUAL SIZE

Figure 3.1   LED and injection laser. (Cross-section views are shown in later illustrations.)

TABLE 3.1 PERFORMANCE TYPICAL OF OPTICAL SOURCES [1]

	LEDs			Lasers			
Wavelength	Class	800–850	1300 and 1550	Class	800–850	1300	1500
Material		GaAlAs	GaInAsP		GaAlAs	GaInAsP	GaInAsP
Spectrum width (nm)		30–60	50–150	MM	1–2	2–5	2–10
Mode spacing (nm)				MM	0.3	0.9	0.13
Line width SM (MHz)				SM FP		150	150
				DFB		10–30	10–30
				EC		1–10	1–10
				EG		0.002–1	0.002–1
Structure				PI, CSP, BH		BH	BH
Output power (mW)		0.5–4.0	0.4–0.6	CSP	20–50		
				BH	2–8	1.5–8	1.5–8
Coupled power (mW)							
100-μm core	Surface	0.1–1.5		MM		0.5–7.0	
	Edge	0.3–0.45	0.04–0.075				
50-μm core	Surface	0.01–0.05	0.015–0.035	MM	0.5–2.5	0.4–3.0	0.5
	Edge	0.05–0.13	0.03–0.06	SM	1.5–3		
Single mode	Edge		0.003–0.03	MM		0.25–1	0.25–0.8
Extinction ratio					25 : 1	25 : 1	25 : 1
Drive current (ma)		50–150	100–150	CSP	40–80		
				BH	10–40	25–130	
Rise time (ns)	Surface	4–14					
	Edge	2–10	2.5–10	BH	0.3–1	0.3–0.7	0.3–0.7
Modulation frequency (GHz)		0.08–0.15	0.05–0.3	BH	2–3	2–4	2–3
Temperature drift							
Wavelength nm/°C		0.3	0.6		0.15–0.2	0.3	0.9
Power, %/°C		−0.2, −0.5	−0.9				
Threshold current, %/°C					0.8	1.6–2	3
Linearity							
2nd harmonic		−30 to −40 dB @ $M_o = 0.5$			−40 to −55 dB		
3rd harmonic		−35 to −40 dB @ $M_o = 0.5$			−50 to −70 dB		
Beam width (half)							
Parallel	Surface	120–180°		CSP	5°		
Perpendicular	Surface	120–180°		CSP	10–25°		
Parallel	Edge	180°		BH	10–25°	10–30°	10–30°
Perpendicular	Edge	30–70°		BH	20–35°	30–40°	30–40°
Lifetime (million hours)		1–10	50–1000		1–10	0.5–50	0.5–50

MM = Multimode
SM = Single mode
FP = Fabry perot
DFB = Distributed feedback
EC = External cavity
EG = External grading
CSP = Channel substrate planar
BH = Buried heterostructure
PI = Proton implantation

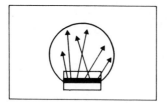 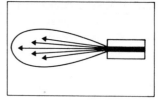

Figure 3.2   Radiating characteristics of an LED (*left*) and an injection laser (*right*). (From Ref. 2; courtesy of AEG-Telefunken.)

3.1.2 Power Output

With present-day ILDs, typically from 0.5 to 5 mW of power is coupled into 50-μm core multimode fiber and from 0.25 to 1 mW into single mode. LEDs, by comparison, couple only about 0.01 to 0.1 mW of power into 50-μm core multimode (about 50 times less). They typically require larger-core fiber to couple more than a milliwatt.

Note that in comparing the power output of these devices, the standard reference is 1 mW, which is generally expressed as 0 dB, that is, 0 dB above the standard reference level of 1 mW. For example, a 0.1-mW LED would be expressed in dB as −10 dB, or 10 dB below 1 mW.

3.1.3 Relative Advantages and Disadvantages

Although ILDs have an advantage over LEDs in coupled power and other characteristics important to high-speed long-distance operation, they have several disadvantages that make LEDs the device of choice for the shorter, slower-speed applications.

1. Cost is much higher for lasers than LEDs for several reasons. The laser is more difficult to manufacture and has more critical optical coupling and heat-sinking requirements.

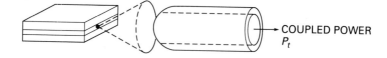

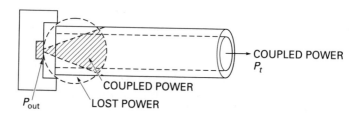

Figure 3.3   Relationship between source power and coupled power [1].

2. Lasers are very temperature sensitive and quickly stop lasing if the chip temperature increases. Thermoelectric coolers or compensating bias control using optical feedback is needed to maintain stable operation, thus increasing the cost of the transmitter circuitry, and packaging and decreasing reliability.

3. The lifetime of laser diodes at room temperature is much less than that of LEDs, largely because of the fact that the current density in the active region is so much higher to create the lasing action.

3.1.4 Lifetime

Lifetimes of LEDs are on the order of 10^7 to 10^8 hours [3], while that for ILDs are on the order of 10^6 hours at room temperature [4].

Obviously, none of these devices has been operated for 10^7 hours, as that represents hundreds of years. However, accelerated life tests allow operating life to be predicted. This is done by elevating the temperature to various levels and observing average time to failure. Failure rate is a direct function of temperature, and thus room temperature failure probability becomes predictable from observing failure rate at high temperature. The approach closely follows that used for determining reliability of all semiconductor devices.

In practice it is rare for a LED or ILD to fail often in a system. They generally will outlast other components such as power supplies and complex circuit board assemblies. Because ILDs in particular use thermoelectric coolers to keep them at constant cool temperature (generally below room temperature) they are very reliable.

3.1.5 Transfer Characteristics

In the LED there is nearly a direct proportion—a linear relationship—between current input and light output, as illustrated in Fig. 3.4. With the ILD the light output at low current levels is similar to that of a LED until a certain lasing threshold is reached. At that point the output power increases rapidly and nearly linearly with additional current.

Both LEDs and lasers can be used to transmit signals that either continuously vary the intensity of the output (analog modulation) or turn the device on and off (digital modulation). To analog modulate a LED, amplified signal current simply drives the LED to produce output power proportional to the input current. To analog modulate a laser diode, the diode must be biased at some point past its lasing threshold, and then the signal current applied. Because neither the LED or the ILD are perfectly linear, they are not well suited for direct analog modulation because they can produce unwanted distortion. Some ILDs, however, have been produced that are linear enough to become popular for use in CATV communications applications.

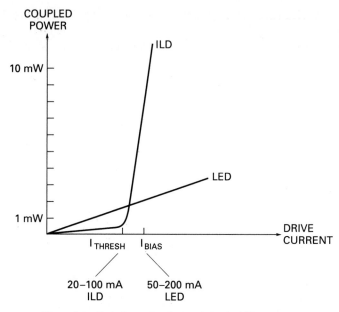

Figure 3.4 Typical transfer characteristics for LEDs and ILDs.

By far most LEDs and ILDs are used for digital communications where the devices are turned on and off by applying current (for a binary "on" state) and then no current (for a binary "off" state). Lasers are generally biased just below threshold for this application.

3.1.6 Response Time

Lasers are faster devices than LEDs with digital modulation of 4.8 Gb/s now possible in certain long-haul transmission products. Fig. 3.5 illustrates the difference in pulse response and frequency characteristics of the LED and ILD. Note that the laser diode is 10 to 100 times faster than the LED.

3.1.7 Wavelength of Operation

Early LEDs and injection lasers transmitted at wavelengths between 815 and 910 nm, that is, from near visible to invisible near-infrared. The GaAs and GaAlAs structures were simple, and relatively mature silicon-detector technology could be used. Also fibers had what was considered at the time to be a low loss attenuation "window" at those wavelengths of around 3 to 8 dB per kilometer.

In time it was determined that fiber would exhibit loss from 0.3 to 1 dB/km at

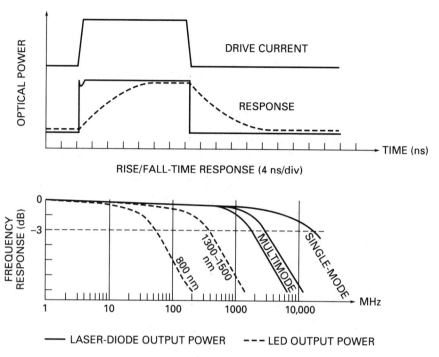

Figure 3.5   Typical response characteristics for LEDs and laser diodes [1].

longer wavelengths (see Chapter 4, Fig. 4.13) so a second generation of devices was developed that operate in the 1300- to 1550-nm range. These devices were made of indium-gallium-arsenide-phosphide (InGaAsP) [5]. Because this wavelength was beyond the range of silicon detectors, germanium was used. Eventually GaInAs detectors were developed because germanium was a very noisy detector material (it had relatively poor sensitivity).

3.1.8 Emission Spectrum

As shown in Fig. 3.6, the spectrum of injection lasers is much narrower than that of LEDs. With broad-spectrum sources, the different wavelengths of light that make up the spectrum will travel down the fiber at different speeds. For high-speed systems, these velocity differences can cause severe spreading of digital pulses, thus reducing the possible modulation rate at which the pulses can be transmitted without interference. This phenomenon, called material dispersion, will be discussed in Chapter 4. The ILD, with its narrower spectrum, causes much less of this effect and therefore is more applicable for high-speed transmission systems.

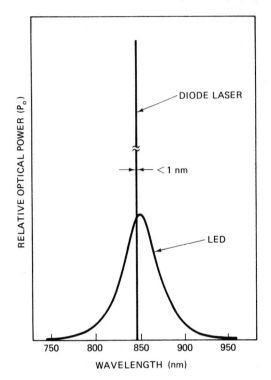

Figure 3.6 Typical emission spectra for injection lasers and LEDs. Typical emission spectra for diode lasers and LEDs show the relatively narrow bandwidth of a DL above the lasing threshold. (Reprinted with permission from Ref. 3 [*Electronic Design,* Vol. 28, No. 8]; copyright Hayden Publishing Co., Inc., 1980.)

3.2 LEDs

Display-type LEDs are well known for their small size, low temperature operation, and ruggedness. Although similar to these LEDs, fiber-optic LEDs are more complex, precise, and carefully made, and they emit more light [6]. Some of the typical symbols for the LED are illustrated in Fig. 3.7.

3.2.1 LED Principles of Operation

As a matter of review, recall that in certain semiconductor diodes, if the *p-n* junction is forward biased, some of the electrons injected across the junction will recombine with holes. If certain materials and dopants are used, a photon will be produced each time an electron falls into a hole, as shown in Fig. 3.8. A ray of light will be formed if a great number of these photons can escape from the diode. The resulting spontaneous radiation is said to be incoherent because the photons are out of step, in random order.

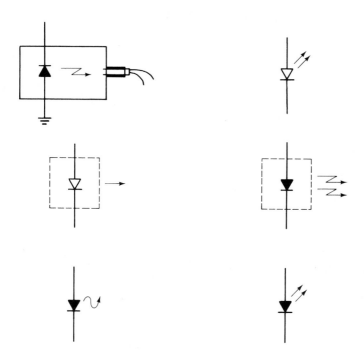

Figure 3.7   Typical symbols for LEDs.

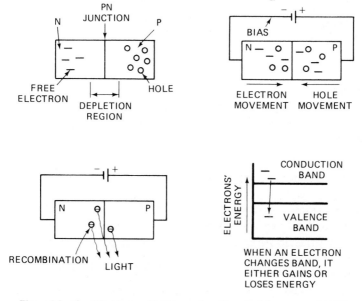

Figure 3.8   Simplified theory of LED operation. (From Ref. 6; courtesy of AMP.)

The materials used in constructing LEDs determine whether the radiation will be visible or invisible (infrared), and if visible, what color. GaAlAs, for instance, will produce invisible (infrared) radiation, whereas GaAsP will generate visible (red) radiation.

Which type should be used depends on the optical fiber and the light receiver to be used. Optical fibers have less attenuation for some wavelengths than others, and light receivers or detectors are more sensitive to some wavelengths than to others.

The light power of the LED is approximately proportional to the injection current, but the LED is never 100% efficient. This is because in some electron-hole recombinations there are no photons produced.

LEDs can be modulated by varying the forward current. Some of the best can be modulated up to 200 MHz.

In comparison with lasers, LEDs have good transient and overload protection [5, p. 54].

Two types of LEDs are in common use in fiber optics: surface emitters (developed at Bell Laboratories) and edge emitters (developed at RCA). Surface emitters are more commonly used, primarily because they give better light emission. However, coupling losses are greater with surface emitters and they have lower modulation bandwidths than edge emitters. Both types are discussed in the following paragraphs.

3.2.2 Surface Emitters

The most efficient surface-emitting LED is the Burrus type, named after its developer. As shown in Fig. 3.9, a well has been etched through the GaAs substrate. This helps to prevent absorption of the radiation. At the same time this well provides a convenient way to bring an optical fiber close to the grainsize light source.

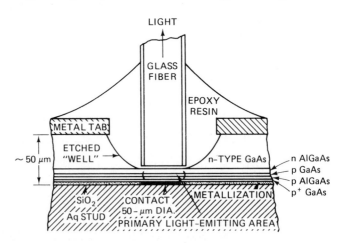

Figure 3.9   Structure of a Burrus surface-emitting LED. (From Ref. 7; copyright 1978 IEEE.)

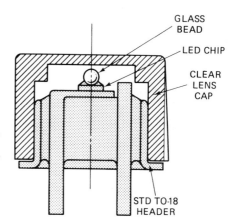

Figure 3.10 Microlens on surface of an LED. A microlens on the surface of the diodes focuses light on the plastic window of Spectronics emitters and detectors. (Reprinted with permission from Ref. 8 [*Electronic Design*, Vol. 28, No. 11]; copyright Hayden Publishing Co., Inc., 1980.)

Because this LED emits light in many directions, it is sometimes necessary to take extra steps to obtain more efficient coupling between the LED and the fiber. This can be accomplished with a microlens (Fig. 3.10) on the surface of the LED chip or a hemispherical dome (Fig. 3.11). These structures are more expensive than the simple surface emitter but allow more power to be coupled into the optical fibers.

Short (12-in) lengths of fibers called *pigtails* are sometimes bonded to LEDs (and injection lasers) to aid coupling. The pigtail is attached to the light-emitting surface. Once installed, the pigtail allows fiber-to-fiber splicing, which can be simpler than aligning the pigtail to the tiny light spot on the LED.

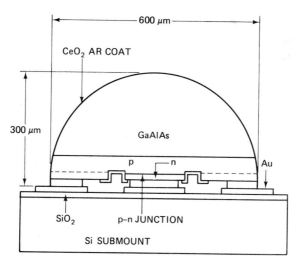

Figure 3.11 Hemispherical dome on LED. The high-power infrared emitting diode from Hitachi uses a forward-biased *p-n* junction to push photons into the GaAlAs crystal dome. The concentration of Al controls the wavelength, and promotes strong emissions. (Reprinted with permission from Ref. 5, p. 59 [*Electronic Design*, Vol. 28, No. 2]; copyright Hayden Publishing Co., Inc., 1980.)

By operating LEDs well below their rated power level—for example, at 20% of maximum drive—their lifetime will go up dramatically, in some cases by a factor of 10 [5, p. 62].

3.2.3 Edge Emitters

As shown in Fig. 3.12, the radiant output of an edge emitter is emitted from the edges of the diode in the recombination region of the junction. An oxide isolated metalliza-tion stripe constricts the current flow through the recombination region to the area of the junction directly below the stripe contact. It is possible to confine the radiating portion of the junction to a spot approximately 50 μm in its greatest dimension. Thus, it provides an excellent match to small-diameter fibers [9, p. 3].

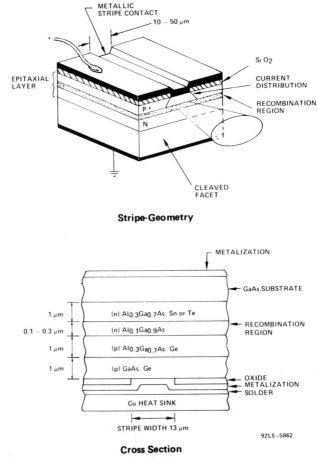

Figure 3.12 Schematic and cross-sectional view of an edge-emitting LED. (From Ref. 9; courtesy of RCA.)

3.2.4 Temperature Considerations

The output power vs. input current characteristics vary with ambient temperature as shown in Fig. 3.13. The variation is somewhat greater in long-wavelength devices.

Heat sinking is used to reduce the amount of self-generated heating. For critical applications the LED can be mounted on a heat sink that is thermoelectrically cooled. This is generally only done for 1300=nm transmission systems, if at all.

3.2.5 LED Transmitters

Figure 3.14 illustrates the basic functional blocks that are required for LED analog and digital transmitters [1].

A LED requires a current source as a driver. For analog transmitters the drive current through the LED must be directly proportional to the signal voltage at the input to the transmitter. In the simplest analog driver in Fig. 3.14(a), the LED is inserted in series with the collector of a transistor, and the current is developed as a function of the voltage across the resistor in the emitter.

For high linearity, high-speed operational amplifiers with feedback make very linear drivers but the inherent nonlinearity of the LED must be compensated for. To achieve this, a diode-matching network is inserted in a feedback arrangement so as to match and compensate for the nonlinear transfer function.

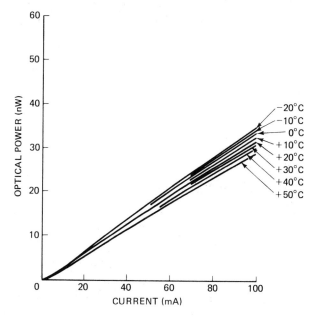

Figure 3.13   Power vs. current characteristics for a GTE laboratory LED. (Reprinted with permission from Ref. 3 [*Electronic Design,* Vol. 28, No. 8]; copyright Hayden Publishing Co., Inc., 1980.)

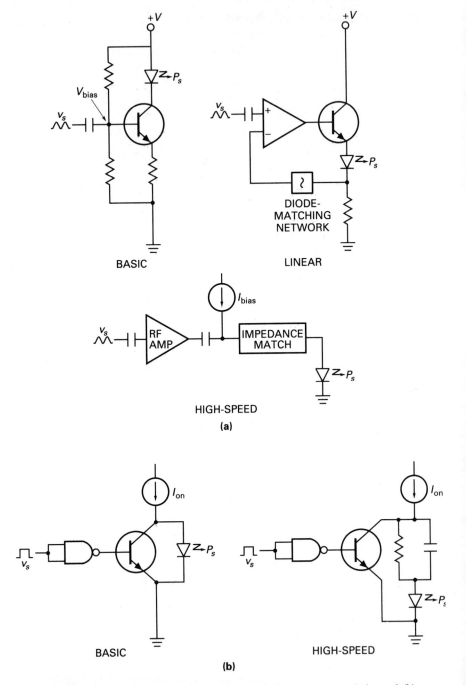

Figure 3.14 LED transmitter-design configurations. (a) Analog transmission and (b) digital transmission. (Reprinted with permission from Ref. 1 [R. Hoss, *Fiber Optic Communications Design Handbook*]; copyright Prentice Hall, Inc., 1990.)

For high-speed operation, above 20 MHz, it is sometimes necessary to compensate for the capacitance and response time of the LED as well. This is generally done by using a resistance and capacitance differentiator network to speed up the response. If 50-ohm high-speed amplifiers are used in the driver, impedance matching is also required.

Digital drivers simply turn the current through the LED on and off. By placing the LED in parallel with the collector-emitter circuit of a transistor, as illustrated in Fig. 3.14(b), the transistor diverts current from the LED (turning it "off") when the transistor is turned on by the binary signal. This driver transistor is preceded by a logic stage (not shown) that provides the standard logical interface and polarity to the transmitting binary signal.

For high-speed operation a resistor-capacitor differentiator circuit is placed in series with the LED to compensate for slow rise and fall time, thus speeding it up.

A LED transmitter designed by Motorola is shown in Fig. 3.15. The transmitter handles nonreturn-to-zero (NRZ) data rates to 10 Mb or square-wave frequencies to 5 MHz, and is TTL- or CMOS-compatible, depending on the circuit selected.

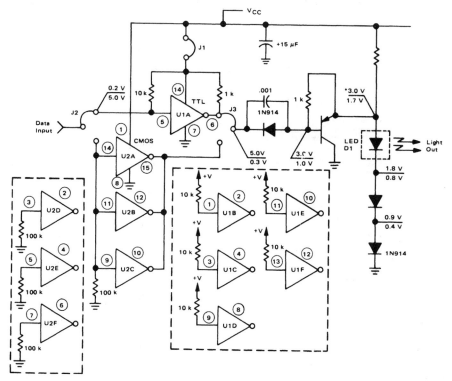

Power Supply: HP6218A or equivalent

Figure 3.15 Fiber optics LED transmitter connected for TTL operation. (From Ref. 10; courtesy of Motorola, Inc.)

Powered from a +5- to +15- V supply for CMOS operation or from a +5-V supply for TTL operation, the transmitter requires only 150 mA of total current. The LED drive current may be adjusted by resistor R1, and should be set for the proper LED power output level needed for system operation. The LED is turned off when transistor Q1 is driven on. Diodes D2 and D3 are used to assure the turn-off. Diode D4 prevents reverse-bias breakdown (base emitter) of transistor Q1 when the integrated circuits U1 or U2 outputs are high [10].

3.3 INJECTION LASERS

Of the three basic types of lasers—gas, solid, and semiconductor— only the semiconductor laser (injection-laser diode [ILD]) is practical for fiber optics. This is because of size, voltage, and cost restrictions.

Although more expensive than a LED, the injection laser, as we have noted earlier, can couple higher power into an optical fiber and is ideally suited to high-speed digital systems (see Fig. 3.16).

Figure 3.16 Injection laser. (Courtesy of RCA.)

3.3.1 ILD Principle of Operation

The injection laser is quite similar to the LED. In fact, they are made of the same materials, although arranged somewhat differently. Below a certain threshold current, the ILD acts as a LED—it has spontaneous emission and a broadband light output. Above the threshold current, the laser starts to oscillate, that is, lasing begins, as shown in Fig. 3.17.

When a properly biased current is passed through the ILD (Fig. 3.18), the holes and electrons move into the active region. Some recombine, giving off photons of light in the process. In the LED the photons can escape the die as emitted light, or they can be reabsorbed by the *p* or *n* material. When a photon is reabsorbed, a free electron may be created or heat may be generated. In the ILD something different happens: The light is partially trapped in the active region by the mirrorlike end walls. The photons reflect back and forth. The photon in the active region, as it reflects back and forth, can persuade a free electron to recombine with a hole. The result is a new photon exactly like the first. That is, the first photon has stimulated the emission of the second. Gain has occurred, for there are two photons where there was but one [6].

For the stimulation to occur, a strong bias current supplying many carriers (holes and free electrons) is required. The current continuously injects carriers into the active region, where the trapped photons stimulate the carriers to recombine and create more photons. The light energy (number of photons) has been pumped up by the carrier injection. This pumping allows amplification [6].

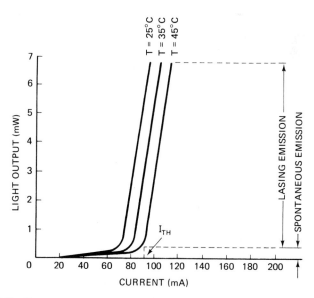

Figure 3.17 Temperature variations vs. light output for injection laser. (From Ref. 11; copyright 1980 Cahners Publishing Co., *EDN*.)

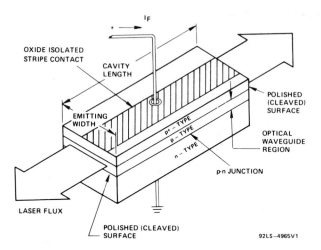

Figure 3.18 Typical injection-laser structure. (From Ref. 9; courtesy of RCA.)

All of the light is not completely trapped in the active region; some is emitted from the mirrorlike end surfaces in a strong narrow beam of laser light [6].

The mounting arrangement of typical ILDs is shown in Figs. 3.19 and 3.20.

3.3.2 Laser-Transmitter Design

The forward-drive current of an ILD must be held at a constant value above the threshold point to maintain a constant radiant flux output. Threshold current, however, is quite temperature dependent and, as a result, the operating temperature must be stabilized to prevent output drifts.

The functional block diagram for laser-diode driver circuitry is illustrated in Fig. 3.21. Note that for both analog and digital modulation the immediate ILD drive electronics is somewhat common in function; only the input signal conditioning is significantly different. The common elements are in the thermoelectric cooler and the bias control.

Figure 3.19 Schematic arrangement of a typical injection laser. (From Ref. 9; courtesy of RCA.)

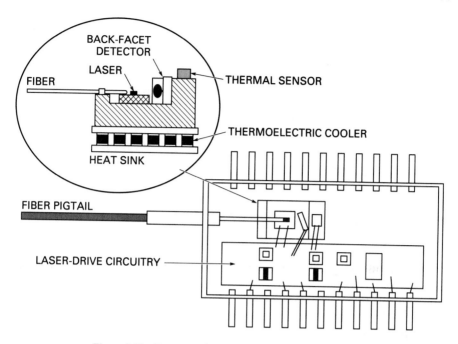

Figure 3.20 Conceptual drawing of a packaged laser transmitter.

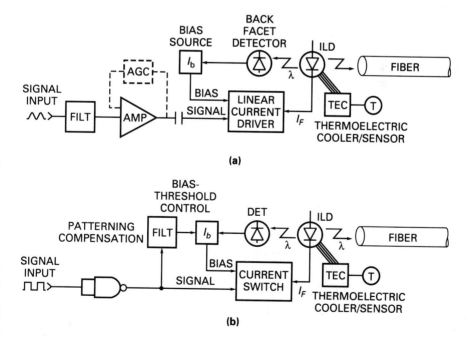

Figure 3.21 Laser-diode transmitter-design configuration. (a) Analog transmission and (b) digital transmission [1].

To remain in the region of the lasing threshold regardless of ambient temperature or internally generated temperature, the laser-diode chip is mounted on a heat sink that is thermoelectrically cooled. A thermistor sensor on the heat sink is in the control loop for the cooler to maintain constant temperature. Generally the temperature is set for a range of 10°C to 20°C. This not only maintains good temperature stability but improves lifetime as well.

A driver-bias circuit incorporating constant current-bias control, and temperature stabilization is shown in Fig. 3.22.

The second important element of the driver, as illustrated in Fig. 3.21, is the optical feedback controlled laser-bias circuit. Based on the principle that laser diodes emit from both the back and forward surfaces, packaged laser modules come with a photodetector that monitors the light output of the back facet (see Fig. 3.20). The photodetector is in a feedback loop to the bias control circuitry. In this way a constant optical output level can be maintained by automatically adjusting current if changes occur because of temperature or operating life.

The bias control circuit also protects the laser from driving beyond the maximum specified radiant flux as well as provides protection from transients. Exposure of the diode to even brief transient current spikes can cause irreversible device failure.

The operation of the circuitry in Fig. 3.21(b) is as follows. The signal input logic circuitry provides isolation between the laser and any signal placed at the input terminal, and a standard logic interface is presented to the input signal.

The current switch transforms the digital voltage waveform from the logic circuitry to a current-modulated signal for the laser. It contains a high-speed transistor stage that supplies the current (in addition to the bias current) to the laser diode that changes its output from just below threshold ("off") to some preset maximum level within the lasing region ("on") as shown in Fig. 3.23.

The variable-current optical-feedback–controlled bias source supplies the DC component of the laser-drive current that maintains the laser diode just below threshold when the signal is in the "off" state or not present. The circuitry contained in this stage is adjusted automatically by the detector monitoring the laser output to maintain the laser output power at either an average value or fixed minimum value (depending on design) regardless of changes owing to temperature, lifetime, or data format and duty cycle. Compensation for signal format and duty cycle is achieved by combining signals from the back facet detector and the input circuitry in an error amplifier to adjust laser bias.

The bias circuit also contains a current limiter so that a maximum bias level is maintained over time and temperature. Once the limit is reached, an alarm is often activated to indicate either laser end of life or some other problem.

Fig. 3.24 illustrates the packaging of a typical laser module with optical pigtail and connector. Table 3.2 illustrates some of the operating parameters for this module [13].

Fig. 3.25 illustrates the block diagram of a lightwave transmitter consisting of a 200-Mb/s data transmitter driven by a balanced ECL signal with a NRZ format. It operates at a 1300-nm wavelength over single-mode fiber. Optical output power is −2 dBm at 50% duty cycle. The InGaAsP injection laser, monitoring photodiode, interface

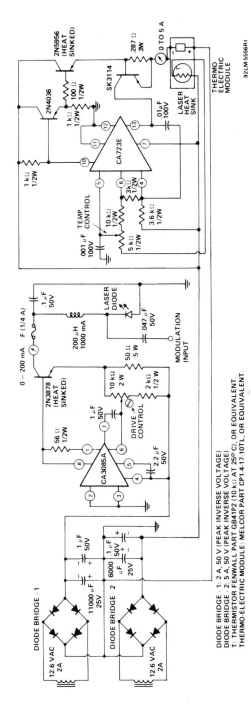

Figure 3.22 Laser-bias circuit with thermo-electric temperature stabilization. (From Ref. 9; courtesy of RCA.)

57

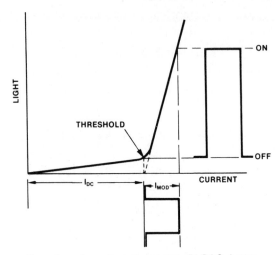

Transfer characteristic of the DI-PAC Laser, the electrical input modulation waveform, and the corresponding light-output waveform.

Figure 3.23   Modulated laser characteristics. (From Ref. 12; courtesy of Optical Information Systems, Exxon Enterprises, Inc.)

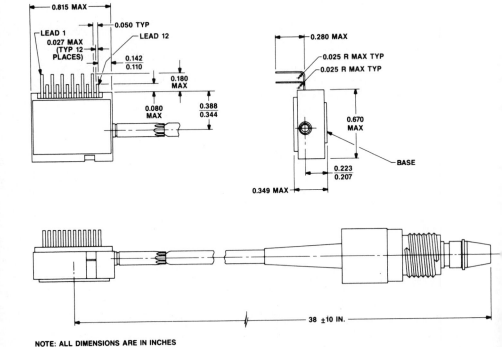

NOTE: ALL DIMENSIONS ARE IN INCHES

Figure 3.24   Thermoelectrically cooled injection-laser module. (From Ref. 13; courtesy of AT&T.)

TABLE 3.2 PERFORMANCE CHARACTERISTICS OF THE ILD MODULE
SHOWN IN FIG. 3.24

USER INFORMATION

Lead Identification Table				
Lead Number	Connection	Lead Number	Connection	
1	Package Ground	7	Monitor (P-Contact) (− Bias)	
2	Package Ground	8	Monitor (N-Contact) (+ Bias)	
3	Thermoelectric Cooler (−)	9	Laser dc Bias (+)	
4	Thermoelectric Cooler (−)	10	Laser dc Bias (−)	
5	Thermoelectric Cooler (+)	11	Temperature Monitor (+)	
6	Thermoelectric Cooler (+)	12	Temperature Monitor (−)	

(Note: See Outline Drawing for pin locations, p. 6.)

CHARACTERISTICS

Electrical Characteristics (T_A = 25°C)

Parameter	Symbol	Min	Typ	Max	Unit
Threshold Current (Figure 1)	I_{TH}	−	18	50	mA
Laser Forward Current for Maximum Optical Output (Figure 1)	I_{FO}	−	38	100	mA
Laser Forward Voltage @ 50 mA (Figure 2)	V_F	−	1.3	1.5	V
Slope Responsivity at $\frac{P_{O(max)}}{2}$ (Figures 1, 2)	R_S	0.02	0.05	0.15	W/A
Maximum Laser Reverse Current	I_R	−	−	100	μA
Monitor Reverse Bias	V_{Mrev}	3.0	5.0	10	V
Monitor Current at $P_{O(max)}$ (Figures 1, 3)	I_{MO}	0.4	1.5	3.0	mA
Monitor Dark Current at I_F =0 and V_{Mrev} =5.0V	I_D	−	0.05	0.5	μA
Laser Submount Temperature at I_{TEC} =0.8 A and $T_{heat\ sink}$ =65°C	T	−	−	25	°C
Temperature Sensor Current	I_{TS}	0.5	−	5.0	mA

(continued)

circuitry, and thermoelectric cooler are contained in an 18-pin hermetic dial in-line package. The laser is thermoelectrically cooled to 20°C. The transmitter operates over a temperature range of 0°C to 50°C. The laser is optically connected to a single-mode fiber pigtail and terminated in an AT&T biconic connector. The transmitter contains hybrid integrated circuits to perform data modulation, laser bias feedback, and protection functions.

TABLE 3.2 *(continued)*

Electrical Characteristics (Continued)

Parameter	Symbol	Min	Typ	Max	Unit
Temperature Sensor Output at I_{TS} = 1.0 mA	V_{TS}	2.94	–	3.02	V
Temperature Sensor Coefficient	dV/dT	9.9	–	10.1	mV/°C
Thermoelectric Cooler Current (Figure 7)	$I_{(TEC)}$	–	–	1.0	A
Laser Junction-to-Temperature Sensor Thermal Impedance	Θ_{JS}	40	–	100	°C/W
Thermoelectric Cooler-to-Heat Sink Thermal Impedance	Θ_{CS}	–	–	4.0	°C/W
Cooler Capacity	ΔT	40	–	–	°C

Optical Characteristics

Parameter	Symbol	Min	Typ	Max	Unit
Wavelength (Figure 6)	λ	1.29	1.30	1.33	μm
Peak Optical Output (CW)	$P_{O(max)}$	−8.0 0.16	−5.0 0.32	−2.0 0.63	dBm mW
Optical Rise Time* (Figures 4, 5) Optical Fall Time*	τ_r τ_f	– –	– –	1.75 1.75	ns ns

* Laser pulsed to P_O = −5 dBm (peak).

Maximum Ratings

Parameter	Rating	Unit
Light Output	−2	dBm
dc Forward Current	100	mA
Laser Reverse Voltage	0	V
Monitor Reverse Bias	10	V
Temperature Sensor Current	5.0	mA
Thermoelectric Cooler Current	1.0	A
Operating Temperature	0 to 65	°C
Storage Temperature	−40 to +85	°C

Source: Ref. 13; courtesy of AT&T.

3.4 MODULATION AND MULTIPLEXING

In designing a fiber-optic system the first step is to select the appropriate modulation and multiplexing approach for the application. Proper choice and design of the modulation and multiplexing largely determines the economies and performance of the

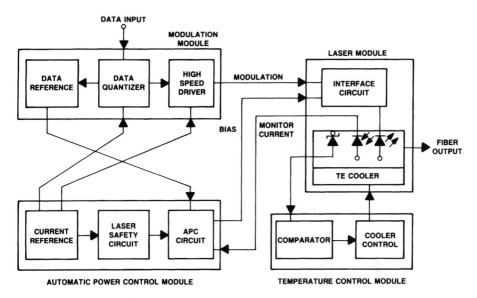

Figure 3.25 AT&T lightwave transmitter block diagram, 1210B. (From Ref. 14; courtesy of AT&T. All rights reserved, printed with permission.)

system. An excellent aid in this process is given in reference [1]. A summary of the approaches and tradeoffs is given here.

3.4.1 Carrier Modulation

Modulation is a process whereby a signal, instead of being transmitted in its original form, is transmitted as amplitude, frequency, or phase changes on another (usually sinusoidal) waveform called a "carrier." The process by which the original signal varies the carrier in amplitude, frequency, or phase is called carrier modulation.

Modulation of a sinusoidal carrier is used in all radio systems today, primarily to convert the transmitted signal to a frequency band where the receivers can best detect it and to separate the different information into frequency channels so that simultaneous transmission can occur. The frequency of the carrier is essentially what one tunes to in a television channel or radio station. Once the carrier is received, the signal can be recovered by observing and detecting the amplitude, frequency, or phase changes that the signal imposed on it. The type of transmission that uses this form of modulation is referred to as analog transmission.

When a signal changes a carrier in amplitude it is called amplitude modulation (AM). When the signal modifies the frequency of a sinusoidal carrier it is called frequency modulation (FM). Likewise a phase variation in a sinusoidal carrier is called phase modulation (PM).

Transmitting a signal by modulating a carrier is done for several reasons.

1. With some modulation approaches such as FM and PM there is a way of trading off necessary received signal power requirements for bandwidth in circumstances such as radio and fiber optics transmission where signal power is low. By adjusting the modulation and demodulation parameters so that a smaller signal level is detectable at the receiver, more bandwidth will be required. With fiber optics, however, bandwidth is less of a problem than power so this is an appropriate tradeoff.

2. Having different channels of information modulate different carriers, each carrier being at a different frequency, permits the channels to be transmitted simultaneously and separated by frequency at the receiver. This is known as frequency division multiplexing and will be discussed subsequently.

3. Modulation of a carrier is sometimes needed to make a signal common in format with other signals being transmitted simultaneously over a particular medium.

Carrier modulation is rarely used with fiber optics, generally when analog transmission offers some advantage in cost or signal format for a particular application. It requires that the optical source be operated in a continuous linear fashion whereby the optical output has a direct linear relationship with the composite modulated carrier waveform.

Fig. 3.26 illustrates the process of carrier modulation in a fiber-optic transmitter. FM modulation is used in this example. The analog signal at the input could represent a voice signal, a video signal, or some other continuously varying amplitude waveform. This signal enters the FM modulator, which is in many cases a voltage-controlled oscillator that produces a fixed amplitude carrier sinusoid at a fixed frequency when there is zero voltage at the input. As signal amplitude increases it causes the frequency of the oscillator to increase and as the signal amplitude decreases it causes the frequency of the oscillator to decrease. The amplitude of the sinusoidal carrier from the oscillator remains fixed.

The resultant sinusoidal carrier with frequency variations is fed to the optical driver which converts it to an equivalent current waveform to drive the LED or ILD. The current waveform directly varies the amplitude of the optical power from the

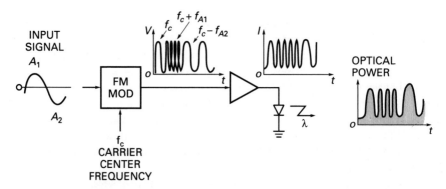

Figure 3.26   Modulation of an optical source by an analog signal.

source as a function of the amplitude change in the carrier waveform. The net result is a sinusoidal increase and decrease in optical power with a fixed minimum and maximum level as dictated by the fixed peak-to-peak level of the carrier. The carrier frequency variations caused by the original signal are imposed on the optical power waveform as variations in the frequency of the sinusoidal peaks as illustrated in Fig. 3.26.

At the receiver the optical power is detected, that is, converted to a current that represents the modulated carrier. The frequency change from the normal carrier zero-state frequency is measured by the demodulator, and the signal is reconstructed.

One case in which analog transmission is desired for fiber optics is in CATV trunking applications where the low cost of AM and FM modulators, in comparison to digital pulse code modulators, outweighs any of the disadvantages in signal performance.

3.4.2 PCM

PCM is the approach most often used to convert an analog signal waveform, such as produced directly by the voice, to a digital on/off waveform, compatible with digital transmission and processing systems. The discussion of PCM will center around one of its biggest applications, the encoding of speech for telephone transmission.

In PCM, a voice (analog) signal is sampled and converted into a binary code of digital pulses. According to communication theory, a voice can be reconstructed if it is sampled at twice its highest frequency. Because the high end of the speaking voice is 4000 Hz, it must be sampled 8000 times per second. The sampling looks at the amplitude of the voice, and expresses the relative amplitude as a binary number that can be represented as pulses. This number is an 8-bit word. Using 8 bits means that a total of 256 different amplitudes can be coded into a binary form ($2^8 = 256$). So for each voice, we must take 8000 samples every second, and every sample has 8 bits. To transmit the voice, 64,000 bits/s (8000 samples/s × 8 bits/sample = 64,000 bits/s) are required (Fig. 3.27). If we wish to send two voices over the same line, then double that—128,000 bits/s—are needed.

3.4.3 Optical Source Modulation

Whether analog carrier modulation or pulse code modulation is employed, the optical transmitter (by the very nature of today's LEDs and ILDs) transmits the signal by varying the amplitude of the power output. As was illustrated in Figs. 3.26 and 3.27, the carrier and signal waveform is simply represented by a proportional variation in optical output power.

In future systems, using coherent optics and single-frequency laser sources, the frequency and phase of the sinusoidal laser waveform may be directly modulated by the signal as is done with radio-wave carriers today. Today's laser, however, has too many spectral components (a jumble of individual carriers) that are not single frequency and therefore can only be amplitude modulated as a group.

With both LEDs and ILDs, direct modulation of the light source by varying the current is the easiest approach for transmission rates of about 5 GHz or so. Indirect or

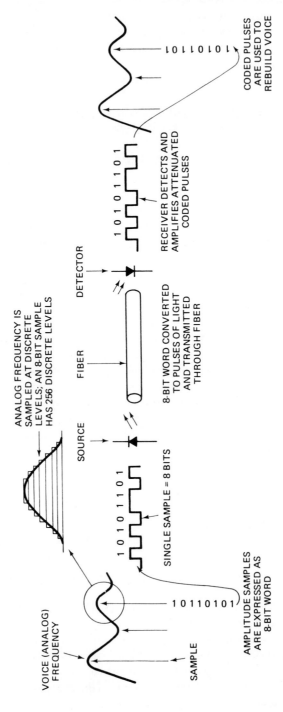

Figure 3.27 PCM. (From Ref. 6; courtesy of AMP.)

external modulation that modifies the light after it leaves the source can be accomplished with electro-optic and magneto-optic modulators. Such modulators are useful for performance beyond 10 GHz [15] and can be built into the substrates of an integrated-optics transmitter. For most present uses, only direct modulation of the source is used.

In digital (pulse) applications, a pulse is formed by turning the source on for a brief instant. The burst of light is the pulse. Digital generally implies a binary signal format or two states: on/off, 1/0, high/low. These two states represent bits or binary digits. Not only does the burst of light have significance as a logic state (binary 1), but its absence has significance as well (a binary 0) as shown in Fig. 3.28.

Fig. 3.27 illustrates the direct modulation of an optical source with a digital waveform, in other words, digital transmission. The specific example shows the conversion of an analog waveform, such as a speech signal, to a digital waveform using PCM. The digital waveform emerges from the pulse-code modulator as a pulse train similar to that in Fig. 3.28. This pulse train enters the optical-source drive electronics where it is converted to a similar current waveform that turns the LED or ILD on and off.

3.4.4 Pulse Encoding

The means by which the information content is represented within a pulse train is known as encoding. PCM is one means of representing an analog signal by encoding the analog samples into a binary representation. Information that is already in binary form can be re-encoded to shape the transmitted pulse train to be compatible with the transmission medium or to permit the optimum use of the bandwidth available. To understand this some terms must be defined.

The following are some of the important terms associated with pulses, as shown in Fig. 3.29 [6]:

1. *Amplitude* is the height of the pulse; it is a measure of the pulse's strength.
2. *Width* is the time the pulse remains at its full amplitude.
3. *Rise time* is the time it takes to go from 10% to 90% of the amplitude. In fiber optics, it is related to how long it takes the source to turn on fully.
4. *Fall time* is the time it takes to go from 90% to 10% of the amplitude. It is related to how long it takes the source to turn off fully.

Rise time is perhaps the single most important characteristic in high-speed digital applications, for it determines how many pulses per second are possible. In the

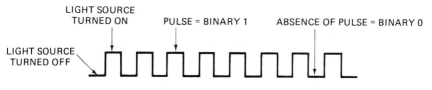

Figure 3.28 Pulse train. (From Ref. 6; courtesy of AMP.)

two pulse trains in Fig. 3.30, the time slots and the pulse widths are identical. The only thing that allows more pulses in one train is faster rise and fall times.

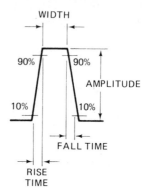

Figure 3.29 Pulse characteristics. (From Ref. 6; courtesy of AMP.)

When we use pulses to represent bits, we speak of speed as a bit rate or so many bits per second. The rising and falling of a pulse is akin to the rising and falling of a sine wave. Bit rate is, in a loose sense, a frequency. Bit rate and sine-wave frequency are related, but they are not the same.

Whereas computer engineers specify transmission channel capacity in bits per second, recall that communications engineers express the same capacity in *bauds*. In some code patterns, the terms are interchangeable; in others, the units must be divided, as will be shown, to obtain equivalent units.

In a NRZ pattern, shown in Fig. 3.31, the signal does not periodically return to zero. The signal will remain at the *1* level if the stream of NRZ data contains a series of consecutive *1*'s. In the same manner, if the stream contains a series of consecutive 0's, the signal will remain at the 0 level. With return-to-zero (RZ) codes, the level periodically changes from high level to low level or back, never remaining at either level for a period longer than 1-bit interval. Because the NRZ code requires only one code interval per bit interval, it uses the channel space most efficiently. Notice that the RZ code uses two code intervals per bit interval [16].

Thus, with NRZ data, 10 mbd corresponds to a data rate of 10 Mb/s. With other codes, the data rate is the baud rate divided by the number of code intervals per bit interval [17].

In general, the analog electrical bandwidth required for an NRZ code is one-

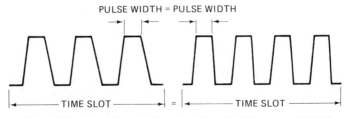

Figure 3.30 Comparison of pulse trains. (From Ref. 6; courtesy of AMP.)

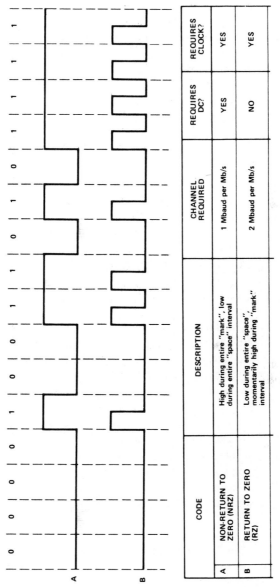

	CODE	DESCRIPTION	CHANNEL REQUIRED	REQUIRES DC?	REQUIRES CLOCK?
A	NON-RETURN TO ZERO (NRZ)	High during entire "mark", low during entire "space" interval	1 Mbaud per Mb/s	YES	YES
B	RETURN TO ZERO (RZ)	Low during entire "space", momentarily high during "mark" interval	2 Mbaud per Mb/s	NO	YES

Figure 3.31 NRZ and RZ code patterns. (From Ref. 16; courtesy of Hewlett-Packard.)

half the bit rate; for an RZ-coded signal, it is equal to the bit rate. Thus, a 200-MHz bandwidth link can transmit 400 Mb/s of NRZ digital data.

A higher bit rate allows more information to be sent over the line. To show why information-carrying capacity is so important, we look at how large amounts of information are sent over a line.

The choice of the RZ or NRZ pattern for fiber-optics transmission is important to the design of the transmitter and receiver. Although NRZ is the most efficient from a bandwidth standpoint, it cannot be transmitted as is through most fiber-optic systems. This is because of the way a receiver must be designed and the way pulses are reconstructed from the received optical signal.

Fig. 3.32 illustrates the process of pulse reconstruction (called regeneration) at the receiver. When the optical signal is detected by the photodetector it is in the form of a very low level (usually less than 1-μA) photocurrent. In order to decode the signal, to determine whether a binary 1 or 0 was sent, the signal must be converted to a voltage and amplified to a level of around 1 to 3 V compatible with the detection logic (referred to as a threshold detector). Practical amplifiers capable of doing this must be capacitively coupled so that the DC bias drift of the first amplifier does not drive the following stages into saturation. Capacitors only pass alternating signals (AC) and block slowly varying or DC signals.

If the received signal is constantly varying, such as might be the case in some RZ coding schemes, it will pass through the capacitors very well because it appears like an AC signal. RZ coding, however, can have long periods where there is no change, such as might be the case with a long series of binary 1's transmitted as shown in Fig. 3.31. This long series of 1's or 0's looks to the capacitor like a DC voltage and is blocked. The result is a signal "droop" on the other side of the capacitor with respect to zero voltage, as illustrated in Fig. 3.32.

The function of the threshold detector is to measure the amplified input signal with respect to standard minimum and maximum voltages and decode anything above a midpoint as a binary 1 and anything below as a binary 0. If the signal droops below this midpoint, threshold decoding errors will occur as illustrated in Fig. 3.32.

Although less serious, the problem is not only with NRZ but exists with RZ as well. As the duty cycle of RZ pulse streams change, the average DC component of the signal changes, causing signal drift up and down at the threshold detector similar to "droop." The solution to the encoding problem is to add a pulse stream encoder to the optical transmitter, which causes the bit states to alter constantly. This is done in two ways.

1. *Phase encoding.* With this encoding the signal is always present as a 50% duty cycle pulse RZ-like pulse stream. A binary 0, for example, may be represented as a pulse at the beginning of a bit period and a binary 1 would be represented by a pulse at the midpoint of the bit period, or such. The disadvantage here is that the signal takes twice as much bandwidth as NRZ; therefore, this approach is usually used with shorter and lower speed systems such as LANs, where transmission bandwidth is not an issue.

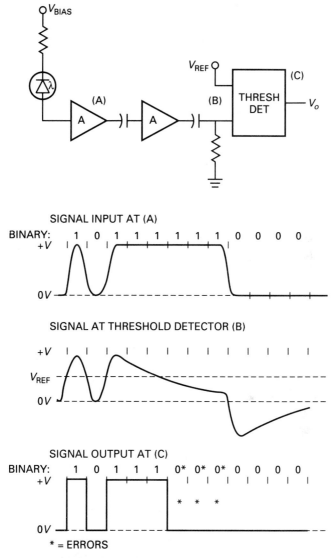

Figure 3.32 Errors owing to AC-coupled signal "droop" at the regenerator with NRZ coding [1].

2. *Scrambling.* A method for preserving the bandwidth efficiency of NRZ encoding, but presenting a near-constant 50% duty cycle signal to the receiver, is to use a scrambling code that in effect inverts every other pulse if a long stream of common pulse states occurs. Various encoding approaches are used to do this. To descramble on the other end the receiver has to know which states are which, so some pulses must be added at the transmitter as a key, which are extracted at the receiver. In time-division multiplexed systems this scrambling and overhead bit insertion-extraction is simply done as a part of the multiplexing process. This approach is used on nearly all high-speed long-distance fiber transmission systems.

3.4.5 Multiplexing

Multiplexing is the means by which we transmit two or more channels of information simultaneously over the same fiber. In fiber optics, three types of multiplexing are common: time-division multiplexing (TDM), frequency-division multiplexing (FDM), and wavelength-division multiplexing (WDM).

3.4.5.1 FDM

FDM is a means of electronically combining multiple channels of information within a single transmission channel by assigning each channel a different carrier frequency. To achieve this, each originating channel (called a baseband channel) amplitude, frequency, or phase modulates a carrier of a different frequency. Each of the new modulated carriers will be referred to as an intermediate channel. Each of these intermediate channels is then combined into a single transmission channel, generally by feeding them into a combiner that is composed of a resistive array (perhaps with an amplifier) not much different from a power splitter one uses for coupling multiple television sets to an antenna.

The result is a composite signal whereby each channel is identified as a separate band of frequencies each identifiable by a discrete carrier frequency. At the receiver the bands or channels can be split using filters and demodulators that are tuned to each of the discrete carrier frequencies.

FDM is useful because it is presently one of the lowest-cost means of combining a lot of channels onto a transmission medium that has a fair amount of bandwidth. This is the reason it is used for radio and television broadcast. Its greatest disadvantage for fiber optics is that the linearity of the optical sources, although some are within the 0.001% to 0.1% range [1], is generally not good enough to avoid the generation of harmonic distortion. These products are essentially mirror images of the signal (or combinations of signals) from one or more channels that appears in some small amount in other channels. The net result is increased noise and crosstalk. To overcome this, more optical power must be added by shortening the transmission distance (increasing source output increases distortion). By reducing the modulation range of the optical source (less optical power) the linearity of the source is generally improved, but here again less power means shorter distances.

Another disadvantage is that fairly sizable "guard bands" are required between channel frequencies in order to prevent interference between channels. This uses up bandwidth, thus limiting the number of channels multiplexed. From 20% to 40% of the total transmission bandwidth can be related to guardband.

The net result is that FDM systems do not perform over the same distance nor with as many multiplexed channels as digitally multiplexed systems do, but the cost tradeoff with today's technology will favor FDM for some applications.

FDM fiber optic systems do not follow any industry standards and are generally manufactured for a specific application such as CATV.

3.4.5.2 TDM

TDM is a means of electronically combining multiple digital channels of information within a single digital transmission channel by assigning each channel a different slot in time to transmit a group of its information bits. TDM is only used with binary signals such as may come from a computer or analog signals that have been digitized using PCM modulation.

To achieve this time-slot assignment, each originating digital channel (1 through N as shown in Fig. 3.33) enters the multiplexer and is held within a buffer memory called a synchronizer. The multiplexer functions like a rotary switch that samples each input channel at a rate greater than N times faster than that of each input channel's data rate. At this rate the multiplexer can sample the first information bit from channels 1 through N, plus add an overhead information and get back to channel 1 before the next bit of information from channel 1 enters.

Depending on the design the multiplexer may accept 1 or more binary bits at a time from each channel and create from it a serial data stream (pulse train) that makes up the single digital transmission signal. The time slotting of channels and overhead information is also illustrated in Fig. 3.33. Overhead bits are used to tell the demultiplexer on the receive end which channel is which so that it can properly separate and reconstruct them.

The advantages of digital transmission and TDM over analog transmission and FDM are

1. Digital transmission requires less optical power at the receiver than does analog; therefore, transmission distances are greater. Transmission distances of 30 to 40 km between repeaters are common for high-speed digital transmission, whereas 10 km to 20 km are typical for broad bandwidth analog.

2. It is difficult to use multiple repeaters with analog transmission because a noise increase and a bandwidth decrease are experienced at the repeater. Digital signals can be repeated hundreds of times.

3. The quality of a received analog signal that has been PCM encoded for digital transmission is not dependent on the quality of the transmission channel, as long as the error rate in decoding the received bits is not too great. With analog transmission the quality or the received signal is directly related to the quality of the transmission channel. As the transmission channel degrades or fades for various reasons, so does the received analog signal. With digitally PCM-encoded analog signals, the quality is almost totally a function of the number of encoding levels used to sample and measure the analog signal, that is, the number of bits per sample (or bits per second) transmitted.

4. TDM multiplexing efficiencies are greater than FDM because no guardband exists with TDM to avoid interference. The only inefficiency in TDM is that a small

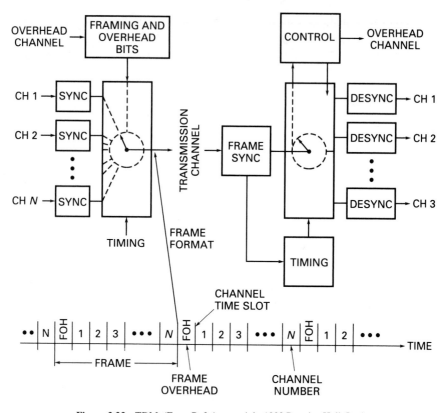

Figure 3.33   TDM. (From Ref. 1; copyright 1990 Prentice Hall, Inc.)

number of bits are added to the transmitted data stream to allow for multiplexer and demultiplexer synchronization and error detection as well as a few extra bits for network management communications. These typically account for less than 10% of the total bits transmitted.

5. There are many TDM and digital interface standards in existence, making interconnection of systems and networks, as well as interworking of equipment from different manufacturers, much easier. Chapter 7 discusses many of the multiplexing and digital interface standards that exist.

The major disadvantage with TDM is the cost of digital PCM encoding. The cost is expected to diminish, however, and the economies will favor digital PCM encoding almost universally.

3.4.5.3 WDM

TDM and FDM are techniques for combining channels, performed on the electrical signals before they enter the optical transmitter. WDM combines multiple channels of light onto a single fiber, using multiple sources at various wavelengths. It is like FDM within the infrared portion of the electromagnetic spectrum. Each optical carrier, at a different wavelength, can carry multiple electrical channels that have been combined using FDM or TDM approaches. WDM, therefore, offers another level of multiplexing to fiber-optic systems that purely electrical systems do not have. To combine the optical signals and separate them at the receiving end, special couplers are used. These WDM couplers are discussed in Chapter 5 (Section 5.3).

REFERENCES

1. R. Hoss, *Fiber Optic Communications Design Handbook* (Englewood Cliffs, N.J.: Prentice Hall, Inc., 1990), pp. 186–244.
2. "Fiber-Optic Transmission Technology, Components and Systems," AEG-Telefunken brochure, p. 6, n.d.
3. R. B. Lauer and J. Schlafer, "LEDs or DLs: Which Light Source Shines Brightest in Fiber-Optic Telecomm Systems?" *Electronic Design,* Vol. 28, No. 8, Apr. 12, 1980, p. 131.
4. "Semiconductor Lasers and Fibre-Optic Components," Philips Component's Division brochure, 1989, Philips International Business Relations, Box 218, 5600 M D Eindhoven, The Netherlands.
5. Stephan Orr, "Fiber-Optic Semis Carve Out Wider Infrared Territory," *Electronic Design,* Vol. 28, No. 2, Jan. 18, 1980.
6. "Introduction to Fiber Optics and AMP Fiber-Optic Products," HB 5444, AMP Incorporated, n.d.
7. S. D. Personick, "Fiber Optic Communication: A Technology Coming of Age," *IEEE Communications Society Magazine,* Mar. 1978, p. 15.
8. Gary M. Null, Julius Uradnisheck, and Ronald L. McCartney, "Three Technologies Forge a Better Fiber-Optic Link," *Electronic Design,* Vol. 28, No. 11, May 24, 1980, p. 68.
9. "Optical Communications Products," RCA Publication OPT-115, June 1979, pp. 12, 13.
10. "Basic Experimental Fiber Optic Systems," Motorola Advance Information.
11. Joseph F. Svacek, "Transmitter Feedback Techniques Stabilize Laser-Diode Output," *EDN,* Mar. 5, 1980, p. 107.
12. "ECL Compatible Fiber-Optic Transmitter Using a Semiconductor Diode Laser OTX5100 Series," Optical Information Systems, Exxon Enterprises, Inc., Preliminary Data Sheet, Nov. 1979.
13. AT&T data sheet, 212A ASTROTEC(TM) Thermoelectrically Cooled Injection Laser Module, AT&T Technologies Inc., 1 Oak Way, Rm 2WC-106, Berkeley Heights, N.J. 07922.
14. AT&T data sheet, "1210A/B Lightwave Transmitters," AT&T Technologies Inc., July 1985.
15. Douglas Lockie, "I Need More Bandwidth!" *Electro- Optical Systems Design,* May 1980, p. 50.
16. "Digital Data Transmission with the HP Fiber Optic System," Hewlett-Packard Application Note 1000, Nov. 1978.
17. "Fiber Optic 1000 Metre Digital Transmitter," Hewlett-Packard, Jan. 1980.

4

Optical Fibers

An optical fiber is a thin, flexible thread of transparent plastic or glass that carries visible light or invisible (near-infrared) radiation. Proper choice of this fiber is vital in fiber-optic system design, as the fiber establishes (1) an upper limit to system bandwidth, and (2) transmitter-to-receiver or transmitter-to-repeater spacing.

Fiber splicing and connections are discussed in Chapter 5; fiber testing is discussed in Chapter 8. Details of fiber and fiber cable manufacture are not given as technicians are not expected to make their own fiber-optic cables.

4.1 PHYSICAL DESCRIPTION

4.1.1 Fiber Construction

As shown in Fig. 4.1, an optical fiber consists of a central cylinder or *core* surrounded by a layer of material called the *cladding,* which in turn is covered by a *jacket.* The core transmits the lightwaves; the cladding keeps the lightwaves within the core and provides some strength to the core. The jacket protects the fiber from moisture and abrasion.

The core as well as the cladding is made of either glass or plastic. With these materials three major types of fiber are made: plastic core with plastic cladding, glass core with plastic cladding, and glass core with glass cladding. In the case of plastics, the core can be polystyrene or polymethyl methacrylate; the cladding is generally silicone or Teflon.

The glass is basically silica, commonly found in sand. Silica occupies 26% of

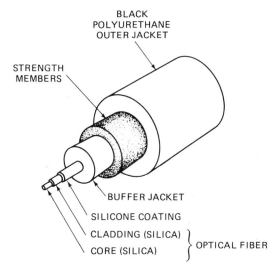

Figure 4.1   Typical optical-fiber construction. (Courtesy of Hewlett-Packard.)

the earth's crust, in stark contrast to copper, which occupies only 0.01% [1]. For optical fibers the silica must be extremely pure; however, very small amounts of dopants such as boron, germanium, or phosphorus may be added to change the refractive index of the fiber. Boron oxide is added to the silica to form borosilicate glass, which is used in some claddings.

In comparison with glass, plastic fibers are flexible and inexpensive. They are easy to install and connect, can withstand greater stresses than glass fibers, and weigh only 40% as much. However, they do not transmit light as efficiently. Because of their high losses, they are used only for short runs such as within buildings. As glass core fibers are so much more widely used than plastic, subsequent references in this book to fibers will be assumed to be glass, rather than plastic, unless specifically stated otherwise.

In comparison to copper, optical fibers are much lighter: a 40-km fiber core weighs only 1 kg; a 1.4-km copper wire of 0.32-mm outer diameter weighs 1 kg [1].

4.1.2 Dimensions

Optical fibers are typically made in lengths of up to 10 km (32,800 ft) without splices. As we will see later, the diameter of the core and the cladding determines many of the optical properties of the fiber. The diameter also determines some of the physical characteristics. The fiber must be large enough to allow splicing or attaching of the connectors. Conversely, if it is too large, it will be too stiff to bend and use too much material (thus becoming expensive).

An optical fiber is very small, comparable in size with the human hair. Its coated diameter is typically between 250 to 500 μm. Compare this with communications grade copper wire, which typically has an outer diameter of from 320 to 1200 μm.

Core diameters range from 8 to 10 μm for single-mode fiber and from 50 to 100 μm for multimode. Outer diameters of the glass fiber are typically held at 125 μm, although 140 μm is used for the 100-μm core fibers. As Fig. 4.1 illustrates, the core is surrounded by a layer of cladding glass. It has a refractive index that is lower than that of the core so as to form an optical waveguide. To maintain most of the light energy within the core, the cladding must have a minimum thickness of one or two wavelengths of the light transmitted [2]. Remaining glass outside of the cladding is for protection and to form a standard fiber size for joining. The outer coating is a plastic material added to protect the glass from surface abrasion. It typically has a diameter of from 250 μm for single coated fibers to 500 μm for fibers with a second (more abrasion resistant) outer buffer. Typical fiber dimensions are given in Fig. 4.2.

Fiber dimensional tolerances must be held very tightly so that fibers can be joined, spliced, and connected with minimal loss. Tolerances for the fiber outer dimensions, and for multimode core diameters, are held within ± 3 μm. Core concentricity and noncircularity are held within 6% [3].

CCITT* recommendation G651 proposes two sets of standard fiber dimensions.

1. For graded index multimode fibers, a core diameter of 50 ± 3 μm and an outer diameter of 125 ± 3 μm
2. For single-mode fibers, a core diameter of 8 μm nominal and an outer diameter of 125 ± 3 μm

4.1.3 Strength

Ordinary glass, as most of us know from experience, is brittle; that is, it is easily broken or cracked. Optical glass fibers, in happy contrast, are surprisingly tough. They have a high tensile strength, that is, ability to withstand hard pulling or stretching. The toughest are as strong as stainless steel wires of the same diameter. In comparison with copper wire, optical fiber has the tensile strength of a copper wire twice as thick.

One-kilometer lengths of these fibers have withstood pulling forces of more than 600,000 lb/in^2 before breaking. Bell Labs reports that a fiber 10 m long (about 33 ft) can be stretched by 70 cm (over 2 ft) and still spring back to its original shape [5]. Yet these fibers can be bent into small radiuses (Fig. 4.3)—as low as 2 mm for a 420-μm-diameter fiber. In demonstrations these fibers have been tied in loose knots (not recommended!) without breaking. Only when the knot was drawn tight did the fiber break.

To produce fibers this tough, fiber manufacturers try to keep the glass core and cladding free from microscopic cracks on the surface or flaws in the interior (Fig. 4.4). When a fiber is under stress, it can break at any one of these flaws.

*Consultative Committee on International Telegraphy and Telephony.

WIDEBAND GRADED INDEX MULTIMODE OPTICAL FIBER
DIMENSIONS SHOWN ARE NOMINAL VALUES

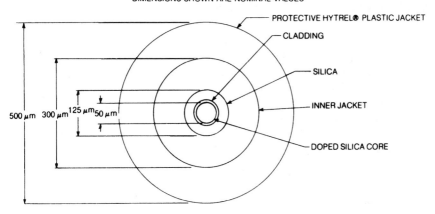

SINGLE MODE STEP INDEX OPTICAL FIBER
DIMENSIONS SHOWN ARE NOMINAL VALUES

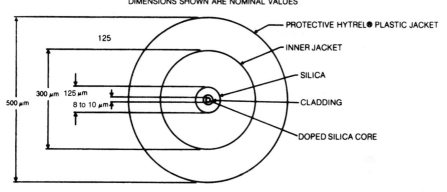

LARGE-CORE PLASTIC-CLAD SILICA OPTICAL FIBER
DIMENSIONS SHOWN ARE NOMINAL VALUES

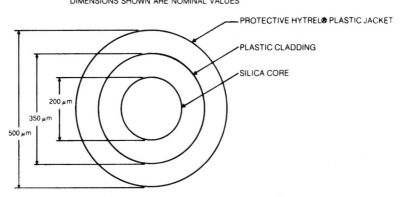

Figure 4.2   Typical optical-fiber dimensions. (Update of Ref. 4; copyright 1978 International Telephone and Telegraph Corp.)

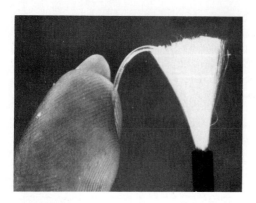

Figure 4.3 Optical fibers can be bent. (Photo courtesy of DuPont.)

Flaws can develop during and after manufacture. Even a tiny particle of dust or a soft piece of Teflon can give the fiber's surface a fatal scratch [8].

To prevent such abrasion, manufacturers coat the fiber with a protective jacket of plastic (organic polymer) immediately after the fiber is made. This jacket also protects the fiber surface from moisture, which can also weaken the fiber. In addition, the jacket cushions the fiber when it is pressed against irregular surfaces. This cushioning reduces the effect of small random bends (microbends) that otherwise would cause transmission losses. Finally, the jacket compensates for some of the contractions and expansions caused by temperature variations.

Although single-fiber cables are used, generally several fibers are placed together in one cable, as described in Section 4.6. Cables are often designed so that there is little or no stress on the fiber itself.

The maximum tensile strength of a fiber may be specified in giganewtons per square meter, newtons, pound-forces per square inch, kilograms, or millipascals. To compare the strength of two fibers, the figures obviously should be in the same units. In addition, the figures should be for the same length of fiber, preferably 1 km.

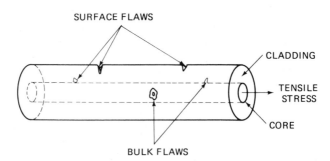

Figure 4.4 Optical-fiber flaws. (From C. K. Kao, "Optical Fibre Cables," in *Optical Fibre Communications,* ed. M. J. Howes and D. V. Morgan; copyright 1980 by John Wiley & Sons Ltd.; reprinted by permission of John Wiley & Sons Ltd.)

4.2 LIGHT PROPAGATION

An optical fiber is produced by forming concentric layers of cladding glass around a core region. The core region maintains the low optical loss properties necessary for the propagation of the optical energy. The higher the refractive index, the slower optical energy will propagate. When a high refractive index glass core is surrounded by the lower-index cladding material (as in step-index fiber), the light energy is contained within the higher-index core owing to the reflection at the interface of the two materials. Furthermore if the refractive index of the core is varied, higher in the center, and lower toward the outside (as in graded index fiber), then the light will be refracted so as to remain within the core.

4.2.1 Refraction and Reflection

Optical fibers guide light by either reflection or refraction, depending on the type of fiber being used. In the reflective types, light rays travel in a zigzag fashion as shown in Fig. 4.5. In the refractive types, light rays travel in a continuous curve (Fig. 4.5). In either case, light rays are confined to the core.

Two reflective types are available: the single-mode step-index fiber and the multimode step-index fiber. Only one type of refractive fiber is available: the multimode graded index fiber.

For the physicist, *mode* is a complex mathematical and physical concept describing the propagation of electromagnetic waves. But for our purposes, mode is simply the various paths light can take in a fiber [17]. By *single mode* we mean that there is only one path for the light; *multimode* means several paths.

To explain *step-index* and *graded-index fibers,* recall that transparent materials have a refractive index (index of refraction). This optical parameter is designated n and can be computed from the equation $n = c/v$. In this equation, c is the speed of light in a vacuum and v is the speed of light in the material.

From this equation we see that a higher index corresponds to a slower speed of light in the material. Thus, light traveling in a material with $n = 1.48$ will travel slower than in a material with $n = 1.41$.

In optical fibers, the refractive index of the core (n_1) is greater than the index of the cladding (n_2). That is, $n_1 > n_2$.

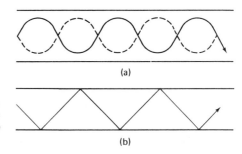

(a)

Figure 4.5 Methods of optical confinement. (a) Refractive confinement and (b) reflective confinement. (From Ref. 6; copyright 1979 T&B/Ansley.)

(b)

In a step-index fiber, the refractive index is constant throughout the core. As shown in Fig. 4.6(a) and (b), the index of the core (n_1) is represented by a flat line, parallel to the horizon. Notice in this profile the abrupt change between the index of the core and the index of the cladding. This abrupt change gives these fibers the name "step index."

In a graded-index fiber, the refractive index is not the same throughout the fiber. It is highest at the center of the core but decreases or tapers off radially toward the outer edge, as shown in Fig. 4.6(c).

In the single-mode step-index fiber, the core is so small that it allows only a single ray to travel down the fiber. This ray in effect travels down the axis of the fiber as shown in Fig. 4.7(c).

Within a step-index fiber, light rays from the core strike the cladding at various angles of incidence, as shown in Fig. 4.8. Ray *A*, which is perpendicular to the inter-

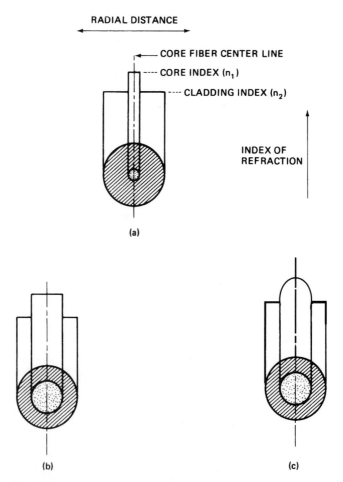

Figure 4.6 Refractive-index profiles. (a) Single-mode step index, (b) multimode step index, and (c) graded index.

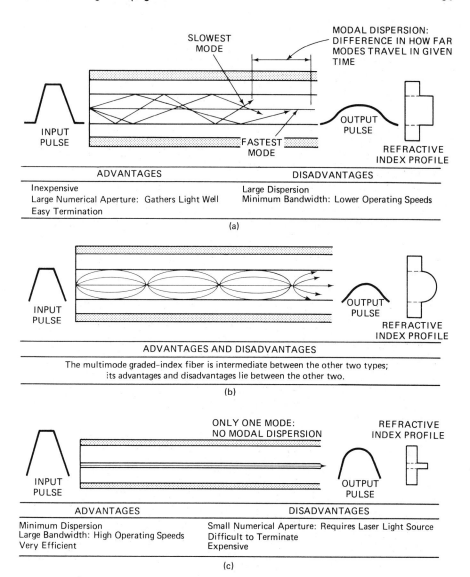

Figure 4.7 Types of light propagation in fibers. (a) Multimode step-index fiber, (b) multimode graded-index fiber, and (c) single-mode step-index fiber. (From Ref. 7; courtesy of AMP.)

face, is transmitted primarily through the interface. Ray *B,* which forms an incident angle φ with the normal, also transmits through the interface, but a larger portion is reflected. As φ becomes larger, more and more of the light is reflected, instead of being transmitted. When φ reaches a certain critical angle φ$_c$, all the light will be reflected. Because of this *total internal reflection,* the light will be confined to the core and will follow a zigzag path (Fig. 4.9) down the core to the other end.

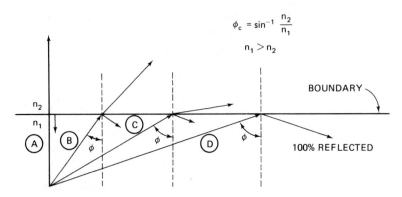

Figure 4.8 Total internal reflection of light rays. For total internal reflection, ϕ must be greater than ϕ_c. (From Ref. 6; copyright 1979 T&B/Ansley.)

From Snell's law (Chapter 2) we can compute the critical angle by

$$\sin \phi_c = \frac{n_2}{n_1}$$

provided that n_1 is greater than n_2 ($n_1 > n_2$). This can be rewritten

$$\phi = \sin^{-1} \frac{n_2}{n_1}$$

For a typical case, $n_1 = 1.48$, $n_2 = 1.46$, and $\phi_c = 80.6$ degrees. Note that total internal reflection will occur only for those rays incident at angles equal to or greater than the critical angle.

The net effect of the critical angle on light coupling or collecting power of the fiber has been defined as numerical aperture (NA). The larger the critical angle within the fiber, the larger the NA. NA is defined as the sine of the half angle of the

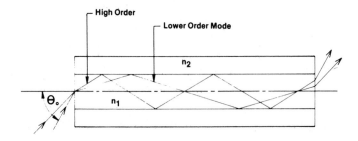

Ray Diagram — Step Index Fiber

Figure 4.9 Refractive confinement in step-index fiber. (From Ref. 6; copyright 1979 T&B/Ansley.)

acceptance cone of the fiber (ϕ) and, like the critical angle, is a function of the refractive indexes of the fiber core and cladding. For multimode fiber (Fig. 4.10),

$$NA = \sin \phi = (n_1{}^2 - n_2{}^2)^{1/2}$$

As illustrated in Fig. 4.11, the larger the NA, the greater will be the amount of light accepted by the fiber. Thus as the NA increases, the greater will be the possible transmission distance, assuming the same light source and detector are used. Unfortunately, as the NA is increased, the bandwidth is decreased, and scattering and absorption losses are increased. The net effect is that large NA fibers are only useful for short-distance low-speed applications. Typically, values for step-index fibers are from 0.2 to 0.5, and graded-index fibers have an NA around 0.2.

In traveling down the optical fiber, each ray of light may be reflected hundreds or thousands of times. The rays reflected at high angles—the high-order modes—must travel a greater distance than the low-angle rays to reach the end of the fiber. Because of this longer distance, the high-angle rays arrive later than the low-angle rays. As a consequence, modulated light pulses broaden as they travel down the fiber. The output pulses then no longer exactly match the input pulses, causing signal distortion. This is discussed in more detail in Section 4.4.

In a graded-index fiber, light rays will travel at different speeds in different parts of the fiber because the refractive index varies throughout the fiber. Near the outer edge, the index is lower; as a result, rays near the outer edge (outer extremity of the core) will travel faster than rays in the center of the core. Because of this higher speed such rays will arrive at the end of the fiber at approximately the same time, even though they took longer paths.

In effect, light rays in these fibers are continually refocused as they travel down the fiber. This refocusing reduces dispersion (to be defined later) and permits operation at much higher data rates.

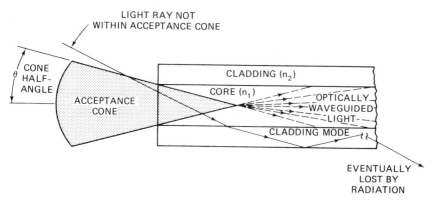

Figure 4.10 Optical fiber's acceptance cone half-angle. (From Ref. 8; copyright 1978 Cahners Publishing Co., *EDN*.)

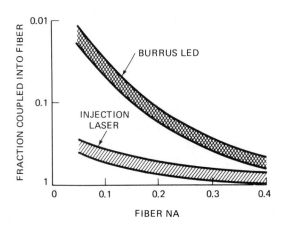

Figure 4.11   Coupling loss vs. fiber numerical aperture. (From Ref. 9; courtesy of General Cable Co.)

Light rays strike the end surface of an optical fiber at many different angles. However, for a ray to be propagated down a fiber, it must enter the end of the fiber within a region called the acceptance cone, shown in Fig. 4.10. That is, a light ray not within the cone will get lost in the cladding and never make its way down the core.

The half-angle of this cone θ is defined for step-index fibers as

$$\sin \theta = n_1 - n_2$$

(*Note:* sin θ is also the NA.) This angle is the maximum angle with respect to the fiber axis at which rays can be accepted for transmission through the fiber. For graded-index fibers, sin θ is the same as the numerical aperture (NA).

Single-mode fiber is a special case of step-index fiber where the core size is made very small (8 to 10 μm) and the refractive index differences between core and cladding are small such that only a single mode of light energy can propagate down the fiber. Because only a single mode or ray of light is traveling down the fiber, the broadening of light pulses owing to different rays of light arriving at the other end at different times cannot happen. Because of the single path through the fiber there is less scattering and absorption of light. Single-mode fiber is therefore the highest-speed and lowest-loss fiber type.

Although the small core size of single mode presents a challenge in coupling light, laser sources and connectors are now designed for excellent coupling with single-mode fiber. Single-mode fiber is sometimes coupled to LEDs, but the loss is extremely high and therefore single mode is almost exclusively used with laser diodes. Splicing to single mode has actually become as good or better quality than with multimode because of the technology of local injection-detection active splicing.

Multimode fiber, having the poorer bandwidth and loss properties, is generally used for shorter applications where the larger core size is needed to couple to lower-cost LEDs. Step-index fiber has the poorest bandwidth performance (less than 10 MHz for a kilometer length) and the highest attenuation (4 to 6 dB/km), but the largest core size for coupling (from 50 to more than 600 μm). It is therefore used only for very short (few hundred meter) local data link or sensor application where LED cost and coupling is important. It is also very large and therefore relatively expensive per unit length because of the amount of material used.

Graded-index fiber has better bandwidth performance (1000 MHz or greater for a kilometer length) and lower attenuation (from 0.7 to 3 dB per kilometer) while retaining a relatively large-core size (50- to 85-μm diameter). It is generally used for moderate bandwidth and distance application, typically 3 to 10 km in lengths at transmission rates in the range of 1 to 150 Mb/s. Intrabuilding optical cable as well as LAN cable is typically multimode-graded index. In comparison with single mode, graded-index fiber uses many more steps in the manufacturing process, is usually purchased in much less volume, and therefore is more expensive per unit length than single mode.

Single-mode fiber with its bandwidth on the order of multiple gigahertz for a kilometer length, and attenuation of 0.3 dB/km is the fiber of choice for high-speed long-haul trunking applications. Typical applications include transmission speeds of 2.4 to 4.8 Gb/s over distances of 30 km without repeaters. It is bought in large quantities and uses the least materials, and therefore is the lowest cost per unit length.

Table 4.1 illustrates the performance characteristics of typical fiber product today [10].

4.3 TRANSMISSION LOSSES

The transmission loss or attenuation of an optical fiber is perhaps the most important characteristic of the fiber, as it generally is the determining factor as to (1) repeater spacing, and (2) the type of optical transmitter and receiver to be used.

As lightwaves travel down an optical fiber, they lose part of their energy because of various light-absorbing compounds and light-scattering mechanisms and imperfections within the fiber. These losses (or attenuations) are measured in decibels per kilometer. A 3-dB/km attenuation, for example, represents a reduction of half of the optical power in 1 km of fiber. For any given cable the attenuation will of course be the fiber unit attenuation (dB/km) multiplied by the length (km) of the cable. The greater the attenuation, the less the light that reaches the detector and thus the shorter the possible distance between repeaters.

Typical low-loss fibers have attenuations of between 0.3 to 3 dB/km. Contrast this attenuation with that shown by coaxial cable in Fig. 4.12. For fibers and coaxial cable alike, the losses are a function of the frequency of the signal carrier. Coax attenuation varies as the square of frequency with signal carriers is in the DC to hundreds of megahertz range. With fiber the usable carrier frequency (band of low attenuation) is in the terahertz range, and therefore we designate optical carrier frequency in terms of its wavelength. Attenuation is therefore specified at certain wavelengths rather

TABLE 4.1 TYPICAL PERFORMANCE CHARACTERISTICS OF CABLED FIBER [10]

General class	Multimode							Single mode		
EIA class	IA and IB				IC	II	III	IVA	IVB	IVC
Index descriptor	Graded and Quasigraded				Step	Step	Step	Dispersion unshifted	Dispersion shifted	Dispersion flat
Core material	Glass				Glass	Plastic	Plastic	Glass	Glass	Glass
Cladding material	Glass				Glass	Glass	Plastic	Glass	Glass	Glass
Profile (g)	1–3 graded, 3–10 quasigraded				>10	>10	>10	N.A.	N.A.	N.A.
Core dia. (μm)	50	62.5	85	100	50–100	200–600	484–980	8.7–10	7–8.7	7–8.7
Clad. dia. (μm)	125	125	125	125–140	125–140	230–650	500–1000	125	125	125
Tolerance										
Core dia.	±3 μm	±3 μm	±3 μm	±4 μm	±8 μm	±10 μm		±8%	±8%	±8%
Concentricity	<6%	<6%	<6%	<6%	<6%	<10%		<1 um	<1um	<1um
Clad dia.	±2μm	±3 μm	±3 μm	±4 μm	±10 μm	±10 μm		±2 μm	±2 μm	±2 μm
Attenuation (dB/km)										
@ 570 nm							70			
@ 650 nm							130–160			
@ 850 nm	2.6–3.5	3.0–4.1	3.0–4.1	3.0–7.0	4.0–6.0	3.0–8.0				
@ 1310 nm	0.7–1.6	0.8–1.8	0.8–1.8	1.5–5.0				0.4–0.7	0.25–0.3	0.4–0.5
@ 1550 nm									0.1	0.25–0.3
@ 2–5 μm predicted	2–5 μm predicted									
Numerical aperature	0.19–0.25	0.27–0.31	0.24–0.3	0.21–0.3	0.15–0.3	0.27–0.37	0.47	N.A.		
Material dispersion (ps/nm/km)										
@ 850 nm	100–120	100–120	100–120	100–120	100–120	100–120		N.A.		
@ 1300 nm	0.9–3.5	3.0–10	3.0–10	3.0–10				0.9–4.0		3.5
@ 1550 nm								20	3.5	3.5
$(BW)_o$ (MHz-km)										
@ 850 nm	200–600	150–500	150–350	20–500	10–60	9–25	0.5	10^{+5}	10^{+5}	10^{+5}
@ 1300 nm	400–1500	300–1000	300–1000	20–400						

N.A., not applicable

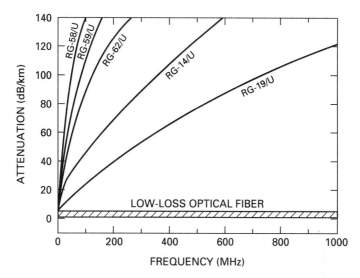

Figure 4.12   Cable attenuation vs. frequency characteristics. (Courtesy of AMP.)

than at certain frequencies. At any one optical-carrier wavelength, the fiber attenuation is nearly unchanged for a broad range of signal frequencies that may amplitude modulate that optical carrier. A typical spectral attenuation plot for fiber product is given in Fig. 4.13.

Note that with the lower-performance step-index fibers, the low attenuation points are between 0.8 to 0.9 μm and 1.0 μm, and the attenuations are on the order of 4 to 6 dB/km. These fibers are used with silicon detectors and GaAlAs sources (usually LEDs). With half of the optical power being lost in about 500 m of fiber, possible transmission distances are short.

With graded-index fibers, operation at 0.85 μm is improved because of a reduction in loss to about 3 dB/km. The real improvement is at 1.3 μm, however, where attenuation is 1 dB/km, and, as seen in the next section, bandwidth is also maximized. Here GaInAsP LEDs and lasers are used along with germanium and InGaAs detectors.

The ultimate in attenuation performance is achieved with single-mode fiber. At 1.3 μm, 0.3 to 0.4 dB/km is achieved, and at 1.55 μm product is available at less than 0.3 dB/km. Note that although reasonable attenuation is measured at 0.85 μm, the fiber is no longer single mode at that wavelength.

Attenuation mechanisms in fiber include the following:

1. Scattering of light because of concentration fluctuations in the dopant oxide causes the greatest attenuation at lower wavelengths [11]. Dopants are used to create the different core and cladding refractive index. Scattering decreases with wavelength to the fourth power, and therefore is the chief motivation for longer wavelength operation.

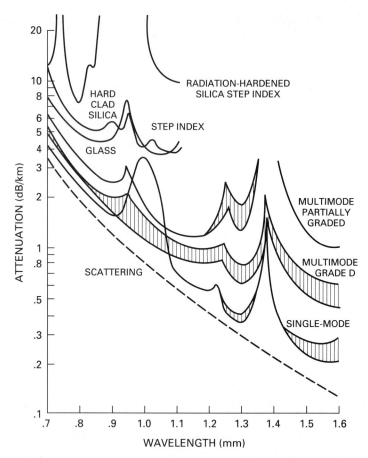

Figure 4.13 Typical fiber-attenuation characteristics. (From Ref. 10; copyright 1990 Prentice Hall, Inc.)

2. Absorption of light by impurities, water, and dopants can be introduced in the glass-making process. These are chiefly responsible for the attenuation peaks and windows in Fig. 4.13.

3. Microbending loss owing to pressure on the fiber distorting the optical waveguide causes light to escape rapidly as seen in Fig. 4.14 [11–13]. This pressure

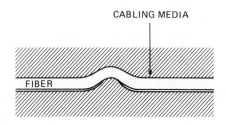

Figure 4.14 Microbending owing to cabling process. (From Ref. 6; copyright 1979 T&B/Ansley.)

is generally a function of the cable construction and fiber-coating materials as well as installation and temperature-related stresses.

4. Core/cladding interface distortion can cause light to escape and is a function of manufacturing quality control or extreme bending of the fiber.

5. Nuclear radiation can cause extremely high transient increases in attenuation as well as more moderate, but sometimes disabling, permanent attenuation increases [14].

6. Absorption of hydrogen into the glass can increase fiber attenuation [14].

4.4 BANDWIDTH AND DISPERSION

Optical fibers, as has been noted earlier, have a higher bandwidth than conventional coaxial or twisted pair cable. For coaxial cable the bandwidth varies inversely as the square of the length [15]. For optical fibers, bandwidth is inversely proportional to length with some differences depending on fiber type. Figure 4.12 illustrated the comparison in performance between fiber and coax.

4.4.1 Bandwidth and Dispersion Relationships

Within most communications systems, *bandwidth* is the term used to describe the performance of the system in terms of signal frequency, that is, the reduction in signal level at the output as signal frequency increases. The frequency at which signal power is reduced by 50% is termed the 3-dB bandwidth, and is usually simply called bandwidth.

Within an optical fiber the mechanism that reduces signal amplitude with frequency is called *dispersion.* It is an effect whereby the colors or modes that make up a wavefront of light are separated as the wavefront travels down the fiber, thus causing it to arrive at the end spread in time.

The difference in width (usually in nanoseconds) between an input pulse and its corresponding output pulse is the pulse dispersion. As it is related to the distance the light traveled in the fiber, it is generally specified per unit length in nanoseconds per kilometer.

If the optical signal is a stream of pulses (a digital signal), the effect is that the pulses widen, often to the point that they will overlap other pulses and smear the information (see Fig. 4.15). This pulse dispersion (or spreading or broadening) makes it difficult for the receiver to tell one pulse from another. It is a form of signal distortion that effectively limits the information-carrying capacity of a fiber-optic system.

The effect on the signal is similar to that imposed by any electronic network that has a limited bandwidth. With some assumptions as to typical pulse shape emanating from the fiber, a relationship between fiber dispersion and equivalent bandwidth can be made [10]. Dispersion is proportional to distance, whereas in nanoseconds per kilometer (ns/km), bandwidth is related as inversely proportional to dis-

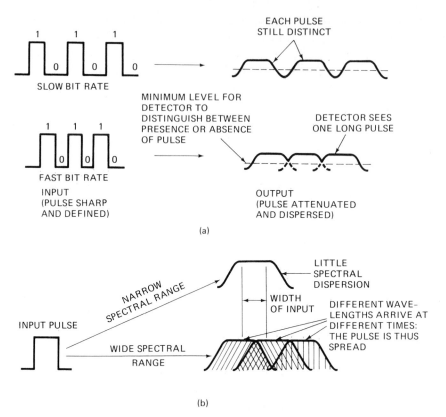

Figure 4.15 Pulse dispersion. (a) Modal dispersion and (b) material dispersion. (From Ref. 7; courtesy of AMP.)

tance in terms of megahertz-kilometer. The relationship between fiber bandwidth and dispersion as seen at the output of the receiver (electrical bandwidth) is

$$\text{Bandwidth (MHz)} = 312/\text{dispersion (ns)}$$

where dispersion is measured as the full width of the pulse spreading at the half-power point of the pulse. This is known as the full-width half-maximum point (FWHM).

Total dispersion is a summation of multimode dispersion and material dispersion.

4.4.2 Multimode Dispersion

Multimode dispersion, also known as intermodal or modal dispersion, is a result of the waveguide geometry and the refractive index differences that permit the fiber to propagate multiple modes or rays of light. In a step-index fiber, for example, light rays that travel parallel to the axis will have a shorter path length than rays that zigzag down the fiber, as shown in Fig. 4.16. Consequently, some rays will take longer to reach the output. For a 1-km fiber with a core refractive index (n_1) of 1.48, a cladding index (n_2) of 1.46, and a core diameter of 50 μm, Kleekamp and Metcalf [8] have calculated that an off-axis light ray incident at an angle of 80.6 degrees would travel 1014 m to reach the end of a fiber 1000 m long. An axial ray in the same fiber would travel only 1000 m. The difference of 14 (1014 minus 1000) means that the off-axis ray would arrive later than the axial ray, even if both rays started at the same instant.

Just how much later?

As we have seen earlier,

$$v = \frac{c}{n_1}$$

Therefore,

$$v = \frac{3 \times 10^8 \text{m/s}}{1.48}$$

$$= 2.03 \times 10^8 \text{ m/s}$$

Since time equals distance divided by velocity, it will take 69 ns (14 m divided by 2.03×10^8) longer for the off-axis ray to arrive.

This time delay is a measure of the pulse dispersion. It can be as low as 0.3 ns/km. If this delay is comparable to the interval between pulses, the output pulses will overlap or spread into adjacent time slots. As a result, the receiver will no longer be able to determine what was sent. In the preceding example, if pulses occurred more frequently than every 69 ns, they would be indistinguishable.

A graded-index core is an attempt to equalize the propagation speeds by reducing the refractive index toward the cladding so as to slow down the rays traveling in the shorter paths so that the longer path rays can arrive at the end at the same time. The shape of the index grading is known as the refractive index profile and is described by the exponent g. Typically g is approximately 2, which represents a para-

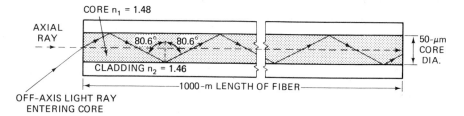

Figure 4.16   Time delay in step-index fiber. *Note:* The off-axis light ray follows a zigzag path 1014 m long, compared with 1000 m for the axial ray. The extra 14 m produces an arrival-time difference of 69 ns. (From Ref. 8; copyright 1978 Cahners Publishing Co., *EDN*.)

bolic profile shape. Step-index fiber has a profile g equal to infinity. Because refractive index is a function of wavelength, graded-index fiber profile can only be optimized for low dispersion at a specific wavelength. *First window fiber* is a term Corning uses to describe fiber optimized for 850-µm wavelength operation. Likewise, *second window fiber* is optimized at 1300-nm wavelength. A *double window* or *broadband fiber* is optimized somewhere in between but has been designed to offer good performance over a broad range of wavelengths and is thus applicable for both 850- and 1300-nm operation.

Single-mode fiber propagates only one mode and therefore has no multimode dispersion effect.

In multimode fiber multimode dispersion characteristics may change with distance. This is known as the *concatenation factor* and is described by the length exponent 8. What happens is that over distance the higher-order modes (long path length rays) either convert to lower-order modes (energy migrates to the core center) or are lost to greater scattering and absorption. The net effect is greater unit bandwidth with distance. The relationship is

$$\text{Total dispersion} = \text{unit dispersion (ns/km)} \times \text{distance (km)}^8$$

Dispersion in step-index fiber, for example, is nearly directly proportional to distance up to about 1 km. At greater distances dispersion becomes proportional to the square root of distance. In other words, the concatenation factor becomes 0.5.

In graded-index fiber the variation with distance is much less than with step index, particularly for the lower-loss 1300-nm fiber. At 850 nm 8 roughly equals 0.7 to 0.9. At 1300 nm 8 is between 0.8 and 1.0, that is, dispersion can be considered directly proportional to distance.

4.4.3 Material Dispersion

Material (also known as chromatic or intramodal) dispersion is caused by the effect of the refractive index of the glass that causes different wavelengths of light to travel down the fiber at different speeds. With a true monochromatic (one-color) light source, there is no material dispersion. The optical source used with fiber optics, however, is a combination of many wavelengths or frequencies of light. The LED in particular has a wide band of wavelengths, typically extending for 30 to 60 nm (0.03 to 0.06 µm) in spectral width. The ILD is much purer in that its spectrum is generally only 2 to 5 nm wide, but even this contains multiple wavelengths that can be separated in time as the light travels down the fiber. Figure 3.6 illustrated the differences in spectral width. Relative differences in source spectral width are directly proportional to differences in material dispersion; thus the ILD gives the fiber greater signal bandwidth carrying properties than does the LED.

Material dispersion is directly proportional to fiber distance as follows:

Total dispersion = unit dispersion (ps/nm/km)

$$\times \text{ source spectral width (ns)} \times \text{ distance (km)}$$

where the unit of dispersion is given in picoseconds of pulse spreading for every nanosecond of source spectral width per every kilometer of distance the light travels.

The characteristics of material dispersion in fused silica fiber are different at different wavelengths. This is illustrated in Fig. 4.17. Note that at a 0.85-µm wavelength the material dispersion is about 120 ps/nm/km, whereas at 1.6 µm it is only about 20 ps/nm/km. Also note that at around 1.3 µm material dispersion is near zero. This is one of the reasons that 1.3 µm is such a popular wavelength for fiber transmission: At this wavelength dispersion is minimal, and thus bandwidth is maximum.

In single-mode fiber, the 1.55-µm wavelength has become popular because of the slightly lower attenuation in comparison with 1.3 µm. To accommodate this, dispersion-shifted single-mode fiber has been developed to produce a near-zero material dispersion at 1.55 µm. Dispersion-flattened fiber also exists, which operates at both 1.3 and 1.55 µm, with low material dispersion.

Note that material dispersion is an important factor in very high-speed, long-

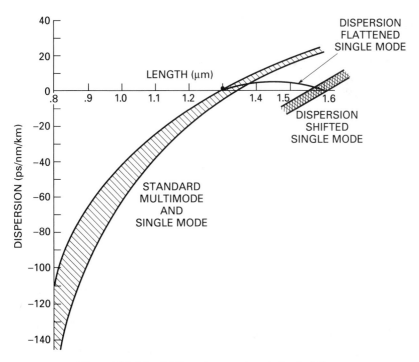

Figure 4.17   Material dispersion range in typical optical fibers.

distance systems. In short step-index fiber using LEDs or moderate-length multimode fiber systems using ILDs, material dispersion is generally negligible in comparison with multimode dispersion. In multimode systems using LEDs or longer multimode systems using ILDs, however, material dispersion is generally of the same order of magnitude as multimode dispersion. In single-mode systems material dispersion is the total bandwidth-limiting mechanism because multimode dispersion does not exist.

To determine the effect of both dispersion mechanisms, multimode (Tm) and chromatic (Tc), they are combined as the square root of the sum of the squares.

$$\text{Total dispersion} = (\text{Tm}^2 + \text{Tc}^2)^{1/2}$$

4.5 CABLE

A fiber-optic cable is one or more optical fibers formed into a cable for convenience and protection. Whether it is buried directly in the ground, hung on telephone poles, pulled through underground ducts, or dropped to the bottom of a lake or ocean, this cable is likely to receive much abuse and mistreatment during its lifetime.

While being installed, it may be stepped on and banged about. Trucks and drums may roll over it. As it is being pulled through ducts, it may be stressed beyond expectation. Once in place it may be subjected to a very cold Canadian winter or a hot Nevada summer. Ice may load it, causing it to sag or break. Gophers and other rodents may try to chew through it. In ice-clogged ducts, technicians may hit it with steam as they clear the ducts. It may be submerged in water in flooded manholes.

The cable must be able to survive this abuse, yet it must be reasonably easy to repair if it breaks, be economically competitive with conventional cables, and be space efficient.

Numerous designs or configurations have been developed to meet these requirements as indicated in Figs. 4.18 through 4.26. These designs differ in materials and arrangements, but practically all of them include coatings to protect individual fibers, strength-bearing materials, filler or buffer materials, and an external protective jacket. In addition, for specific applications some cables include armor protection against rodents and copper wires for carrying electrical power.

4.5.1 Cable Configuration

The optical fiber is coated with soft silicone or similar buffer materials, immediately upon fabrication to prevent damage from abrasion, dirt, and moisture. The coating is considered part of the fiber, not the cable. An additional coating or jacket of a durable plastic or nylon may be added for still further protection. The jacket may be color-coded for easy identification during installation and repair.

The strength or tension member minimizes or eliminates stretching force (tensile stress) applied to the optical fibers. It is also called the load-bearing member. It is

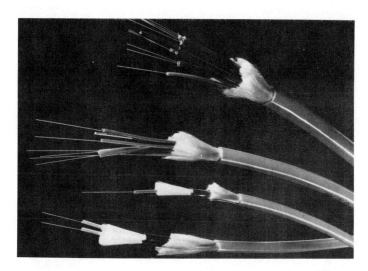

Figure 4.18   Typical cable configurations of Siecor. (Courtesy of Siecor.)

generally made of stranded steel wire, braided Kevlar® aramid yarn, or a fiberglass compound. Both Kevlar and fiberglass are nonconducting, and are therefore used where the cable would be susceptible to lightning damage.

Fiber containing only the primary buffer coating is generally protected by either placing it in rigid slots (usually polyethylene) or within rigid tubes. These slots and tubes are then filled with a gel to keep water out. When the fiber contains a second rigid overcoating, it is often protected by stranding the Kevlar® strength members over the fiber bundle.

The outer protective jacket may be made of polyethylene, polyurethane, fluoropolymer, polyvinyl chloride (PVC), or Tefzel. It protects the fibers from dirt, moisture, sunlight, abrasion, crushing, and temperature variations. Like the individual fibers, it may also be color coded. Length markers and cable type may be imprinted on this jacket. Flame-retardant types are available. In some cases this jacket may carry some of the load, just as the strength members do.

4.5.2 Cable Core Design Classifications

There are as many cable designs as there are cable manufacturers, but some basic cable core configurations have emerged as somewhat common. Three designs predominate among manufacturers: buffer tube design; slotted core design; and tight buffer design. These are illustrated in Fig. 4.19. A fourth design, which is used in variation by manufacturers such as AT&T, is also illustrated in the figure.

Buffer Tube.   In the buffer tube design, illustrated in Figs. 4.20 through 4.22, fibers are placed in plastic tubes (buffer tubes) before being placed into the cable.

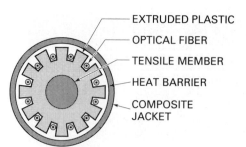

EXTRUDED PLASTIC
OPTICAL FIBER
TENSILE MEMBER
HEAT BARRIER
COMPOSITE JACKET

(A) Slotted Core-Optical cable design in which the fiber runs on a void.

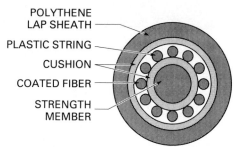

POLYTHENE LAP SHEATH
PLASTIC STRING
CUSHION
COATED FIBER
STRENGTH MEMBER

(B) Cushioned Core-Optical cable design in which the fiber is supported by soft cushions.

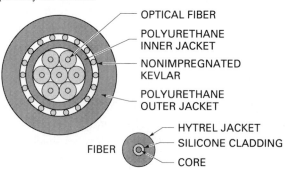

OPTICAL FIBER
POLYURETHANE INNER JACKET
NONIMPREGNATED KEVLAR
POLYURETHANE OUTER JACKET

FIBER
HYTREL JACKET
SILICONE CLADDING
CORE

(C) Tight Buffer-Optical cable incorporating fully supported fibers.

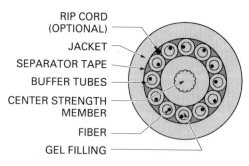

RIP CORD (OPTIONAL)
JACKET
SEPARATOR TAPE
BUFFER TUBES
CENTER STRENGTH MEMBER
FIBER
GEL FILLING

(D) Buffer Tube-Fiber is protected inside gel-filled plastic tubes.

Figure 4.19 Main categories of optical cable design. (From Ref. 17; reproduced by permission of Electrical Communication.)

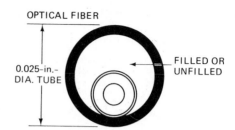

OPTICAL FIBER

0.025-in.-
DIA. TUBE

FILLED OR
UNFILLED

Figure 4.20   Optical fiber in buffer tube. (From Ref. 9; courtesy of General Cable Co.)

When the tube is extruded over the fiber it shrinks slightly, forcing the fiber to spiral inside, therefore, leaving room for the fiber to move if the tube is stretched during installation or afterward because of either tension or temperature change. The fiber is isolated as well from crushing forces by the hard outer tube. The tube is filled with a soft gel to prevent water and dirt from entering. The design is offered by most cable manufacturers.

Slotted Core.   The slotted core design is illustrated in Figs. 4.19(a) and 4.23. Like the buffer tube design it permits fibers to move under tensile forces as well as isolates the fiber from crushing. The design is used extensively by Northern Telecom [16]. In its design the core is made from polyethylene, can have from six to eight slots, and can contain as many as 144 fibers. Northern also uses a patented approach to extruding the slotted core so that the slots oscillate in a spiral-like fashion. This gives the core and fiber room to straighten out if the core is under tension, thus isolating the fiber from tensile forces.

Tight Buffer.   In the tight bound or tight buffered design, Fig. 4.19(c), the main protection against crushing of the fibers is the hard buffer coating applied directly over the primary soft buffer coating of the fibers. Because there is no "window" for fiber movement the tensile forces must be taken up in other ways. This is generally done by stranding the buffered fiber in a spiral fashion in the center of the

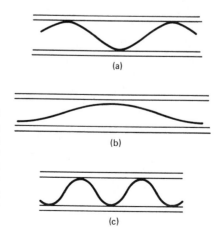

(a)

(b)

(c)

Figure 4.21   Optical fiber with excess length inside a loose buffer jacket. (a) Fiber in buffer jacket after cable manufacturing, (b) decrease of fiber excess length caused by strain of buffer jacket during cable stress, and (c) increase of fiber excess length caused by shrinkage of buffer jacket materials during cooling. (From Ref. 18; courtesy of Siecor Optical Cable and Siemens AG.)

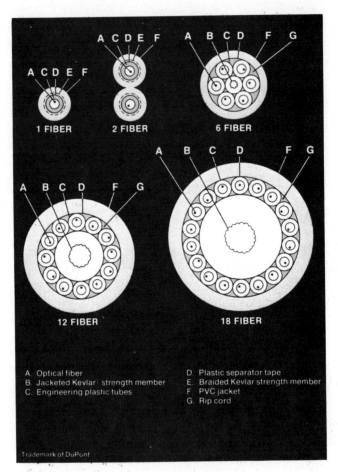

Figure 4.22 Typical cable configurations with loose fit in buffer tubes. (Courtesy of Belden Corporation.)

cable core and surrounding it with Kevlar® or Aramid yarn. If the cable is placed under tension the stress is almost totally taken up by the Kevlar or Aramid yarn.

This fiber design forms the basis for intrabuilding cable and optical pigtails and jumper cables, because the rugged fiber buffering makes individual fibers easy to handle, and it permits small, very rugged cables to be fabricated.

In a demonstration of a ruggedness of army field cable system built by Mitre Corporation for the air force, tight-bound cable did not break after 30,000 vehicle crossings, whereas a buffer tube design exhibited fiber breakage after only 500 crossings [19]. The field cable is still produced by Optical Cable Corporation.

The advantage of the tight buffered design is its ruggedness and handling. The disadvantage is poor temperature performance. As temperatures fall below the 10°C to −10°C range (depending on design), the plastic buffering and overcoating materi-

Type N core, 8 slots, Stainless Steel Strengthmember, Polygel Fill, Polysteelpeth Jacket

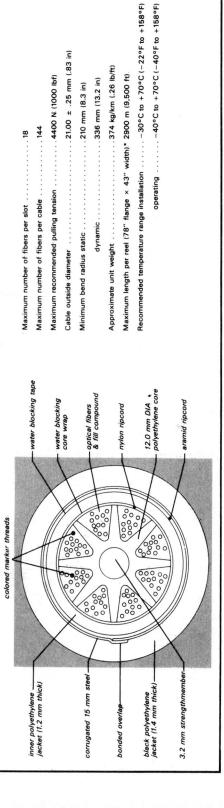

Maximum number of fibers per slot 18
Maximum number of fibers per cable 144
Maximum recommended pulling tension . . . . 4400 N (1000 lbf)
Cable outside diameter . . . . 21.00 ± .25 mm (.83 in)
Minimum bend radius static . . . . 210 mm (8.3 in)
                        dynamic . . . . 336 mm (13.2 in)
Approximate unit weight . . . . 374 kg/km (.26 lb/ft)
Maximum length per reel (78" flange × 43" width)* 2900 m (9,500 ft)
Recommended temperature range installation . . . . . . −30°C to +70°C (−22°F to +158°F)
                        operating . . . . −40°C to +70°C (−40°F to +158°F)

Figure 4.23 Slotted core design. (From Ref. 16; 14050/05 issue 2).

als harden and shrink, causing some small amount of pressure on the fiber. This is enough to create microbending losses, however, that can add another 0.2 to 0.5 dB/km to the attenuation depending on materials and temperature. Below −30°C most tightly buffered fiber is unusable.

Cushioned Core. The design illustrated in Fig. 4.19b protects the fibers by stranding them around a cushioned strength member. The strength member is generally of a fiberglass compound and the cushioning is a soft silicone bonded to the core. The fiber is spiral stranded over the strength member so that under tensile forces the fiber can move while the tension is taken up by the strength member. Fibers are surrounded on the outside by gel filling and a hard outer sheath. The sheath not only protects against crushing but is like a small conduit that forms a water barrier.

Bundled Fiber. The bundled fiber design (not shown) developed by AT&T, protects the fibers by loosely bundling them inside of a gel-filled hard outer sheath, similar to the sheath used in the cushioned core design. Tensile forces are taken up by strength members outside of the hard sheath.

4.5.3 Cable Jacket and Armor Design

Cable jacket and armor are designed for the particular environment that the cable will be installed into. This will be discussed in more detail in the next section. Common jacket and armoring designs are illustrated in Fig. 4.24.

Figure 4.24(a) illustrates designs that are useful where high electric fields exist. Varying degrees of inner jacketing are applied depending on crush resistance required.

Figure 4.24(b) illustrates a jacket, generally of polyethylene, with a single armor wrap of aluminum or steel. The aluminum wrap is only for aerial or duct installation where the risk of rodent damage is minimal. The corrugated steel wrap is used for aerial, duct, and direct burial application. The water-blocking core wrap is used with gel-filled cores. The thicker polyethylene jacket is used for direct burial applications where additional crush strength is required.

Figure 4.24(c) illustrates a double-jacketed design. With corrugated steel wrap it is generally used in direct burial applications where abnormally high levels of stress are expected. Thicker inner jacketing is used where additional crush and lightning protection is desired. It is also used in aerial, burial, or duct applications with a copper wrap when lightning resistance is needed. In this case the strength members are dielectric. In an AT&T direct burial design, copper is wrapped over the steel armor to provide both rodent protection as well as lightning protection [26].

Figure 4.24(d) illustrates the steel-wire armor approach used by AT&T. These cables are applicable for aerial, duct, or burial and have a tensile strength of 600 lb. Rodent protection is added with a steel wrap and lightning protection with a copper wrap.

Armored designs with added tensile members for 1000-lb pull strength are il-

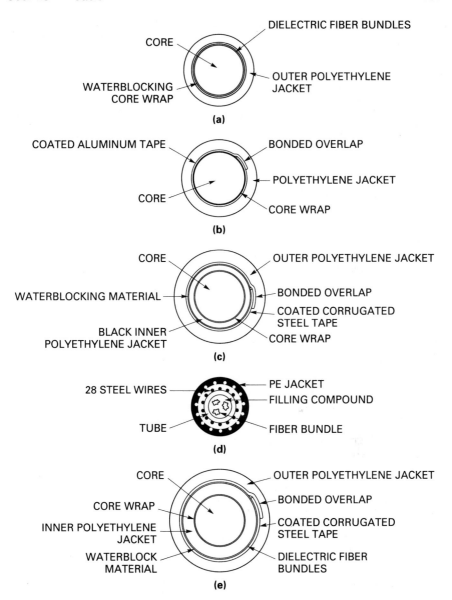

Figure 4.24   Common cable jacket and armor designs. (a) All dielectric cable, (b) single jacketed with aluminum or steel wrap, (c) double jacketed with steel wrap, (d) double armored steel wire, (e) double jacketing with added tensile members. (From Ref. 20 [a–c, e]; copyright Northern Telecom. From Ref. 21 [d]; copyright AT&T 1985.)

lustrated in Fig. 4.24(e). It is recommended for aerial, direct burial, and duct applications where the high tensile forces occur during installation. Lightning protection is enhanced by making the tensile members dielectric and increasing the size of the inner jacket.

4.5.4 Core/Jacket Designs for Outside Plant

Some of the design principles generally used in selection of cables for a particular outside plant installation are given in this section.

4.5.4.1 Cable Design for Aerial Installation

Jacket.   The jacket is of a polyethylene or polyurethane material resistant to water, ultraviolet, and temperature extremes.

Armor.   Steel armor is generally used to provide a degree of rodent protection. Only where rodent damage is known to be negligible should aluminum armor or dielectric armor be used.

Inner Jacket.   An inner jacket of similar material to the outer jacket, or water-blocking core wrap, is recommended to form a water barrier.

Gel Filling.   All air gaps between fiber and cable core elements are filled with a nonhygroscopic gel to prevent water from entering the core elements or contacting the fiber.

Strength Member.   Cable for aerial installation requires strength members that protect against the extreme tensile loads of installation. Some cable can be purchased with integral steel strand wires for hanging [22]; however, this is generally used only for drops to buildings and is not a common practice for long trunk runs. Cable is generally lashed to a preinstalled heavy steel strand that takes most of the tensile load after installation. During installation, however, tensile loads resulting from hanging and pulling can be high. A tensile strength of 600 lb is generally required.

Strength members are generally external steel wire or internal dielectric so that lightning is not attracted through the fiber section of the core. This also eliminates any need for grounding of the cable core materials.

Fiber Core Unit Design.   The core design is generally one of the loose fiber designs (buffer tube or slotted). Care must be taken in using a tight buffer product that has low temperature performance constraints, otherwise extreme cold temperatures will severely affect system performance.

4.5.4.2 Cable Design for Duct Installation

Jacket.   The outer and inner jackets are designed to provide water barriers because the cable will most certainly be under water for extended periods. The jacket is of high-density polyurethanes or polyethylenes so as to be resistant to abrasives, water, chemicals, and gels or grease used in installation process.

Armor.   In city duct, steel armor should be used for rodent and puncture protection. As illustrated in Fig. 4.24, corrugated steel is most common, although steel wire is sometimes used. Corrugated steel wrap must be of a bonded design so that it will not separate when subjected to the rough installation forces.

Where PVC or steel duct has been placed in a trench, and fiber pulled through from pull boxes, an all-dielectric cable is preferred in areas of frequent lightning. An all-dielectric cable does not attract lightning but gives no rodent protection. This may be satisfactory in sealed duct.

Inner Jacket.   Because of the possibility of constant water immersion an inner jacket over the fiber core is recommended to form a water barrier. This is particularly the case where outer jacket and armor may get pin-holed because of lightning strikes.

Gel Filling.   All fiber and all air gaps between core components are filled with a nonhygroscopic gel to prevent water from entering the core elements or contacting the fiber. The gel is of a type that will not flow from the cable under high temperatures ($70°C$ or so) that may exist when ducts cross steam pipes. The gel filling must be specified to withstand a water pressure much greater than the depth of the fiber.

Strength Member.   The strength member can be Kevlar, epoxy-glass, or steel. If the strength member is in the center of the cable core, a dielectric is required in order not to attract lightning. It, and the cable structure, should be able to withstand up to a 600- to 1000-lb pull strength without fiber damage. This is because the pulling winches will exert such tensile force on the cable as it is pulled through duct and around bends.

Fiber Core Unit Design.   The core design is generally one of the loose fiber designs. Care must be taken in using certain tight buffer cables that have low temperature performance constraints.

4.5.4.3 Cable Design for Direct Burial

Table 4.2 and Fig. 4.23 give characteristics typical of outside plant cable. All except the nonmetallic and air core are suitable for direct burial.

Jacket.   The jacket must be water resistant, and particularly resistant to abrasion and puncture and to gels or grease used in installation process.

Armor.   Armor design for direct plow in cable is critical. Because the plow can place a great deal of localized bending and crushing forces on the cable, the armor must be designed to withstand this installation method. Corrugated steel is

TABLE 4.2 CHARACTERISTICS OF OUTSIDE PLANT CABLE

Sheath type	Ribbon Diameter (in)	Weight (lb/ft)	Lightpack Diameter (in)	Weight (lb/ft)	Pulling tension (lb)
Crossply metallic	0.49	0.09	0.42 0.49[a]	0.08 0.12[a]	600 600
Metallic, lightning protection	0.49	0.10	0.42	0.09	600
Rodent/lightning protection	0.49	0.12	0.42 0.49[a]	0.11 0.12[a]	600
Nonmetallic	0.49	0.09	0.43	0.08	600
Metallic/PVC (air core only)	0.49	0.11	NA	NA	NA

NA = Not available in this core/sheath configuration.
[a]Fiber counts 50–72.

Submarine Lightguide Cable Physical Data				
Type of armor	Crossply cable diameter (in)	Armored cable od (in)	Weight (lb/ft)	Approx. tensile rating—short term (lb)
Single-wire armor (swa)	0.49	1.09	1.00	9,000
Double wire armor (dwa)	0.49	1.28	1.86	19,000
Lead plus double-wire armor (pb-dwa)	0.49	1.53	4.30	19,000
Triple-wire armor (twa)	0.49	1.80	3.85	40,000

Source: Ref. 23; courtesy of AT&T.

generally used for rodent protection. The steel wrap must be of a bonded design so that it will not separate when subjected to the rough installation forces.

Lightning can penetrate well under ground. In areas where lightning is common, and the cable direct buried, it is best to use a secondary copper wrap as an electrical conductor. If the fiber is buried within sealed underground steel or plastic duct it is best to use all dielectric cable, which will not attract lightning.

Inner Jacket. An inner jacket over the fiber core is recommended to form a water barrier, and add crush and lightning resistance.

Gel Filling. All fiber and all air gaps between core components are filled with a nonhygroscopic gel to prevent water from entering the core elements or contacting the fiber.

Strength Member.   The strength member may be Kevlar, epoxy-glass, or steel. If the strength member is in the center of the cable core, a dielectric is required in order not to attract lightning. It, and the cable structure, should be able to withstand 600- to 1000-lb tensile load, without fiber damage. The plow will exert tensile force on the cable, and the cable may have to be pulled through underground conduit at end points.

Fiber Core Unit Design.   The core design is generally one of the loose fiber designs. Although underground temperatures are somewhat stable, certain tight buffer designs are less desirable because of possible low temperature extremes in splice pits, incidental duct, and areas where the cable extends above ground.

Precaution must be used in plowing the open channel core design because of the chance that fibers can migrate out of a channel during the strenuous operation and become pinched between the plastic core and the inner jacket. The buffer tube design is not susceptible to this type of damage.

4.5.5 Premise Cable Design

Figure 4.25 illustrates some of the typical cable designs for premise cabling, and lists some of their characteristics [22]. Generally the cable core is of the tight buffered design, although the buffer tube is sometimes used. The cable is almost always all dielectric and uses Kevlar as a strength member.

4.5.5.1 Riser Cable

The outside sheath will be required by National Electrical Code to be fire retardant and possibly of a material that will not emit poisonous gases when exposed to flame. Some codes will permit flammable sheaths if installed in duct. Cables generally meet a UL classification per section 770-6 of the National Electrical Code. Fire-retardant PVC or polyurethane, Teflon, or a fluoropolymer is generally acceptable (Fig. 4.25), although some codes may require other materials.

Riser cable has strength members within it to be able to support its own weight in lengths of about 500 ft without adverse effects on fiber attenuation or lifetime. It also has a reasonable bend radius so that it can be pulled within interior riser duct and shafts. Typical specifications are a 5-in bend radius (10 in under load) and a maximum pulling tension of 600 lb. Installation temperature extremes range from −20°F to +120°F. Weight is about 3 to 10 lb per 100 ft.

The cable core construction and fiber units are generally designed so that individual fibers can be broken out of the cable at multiple points up the riser. For this reason the tight buffered core design works best. Fibers that are individually strength-

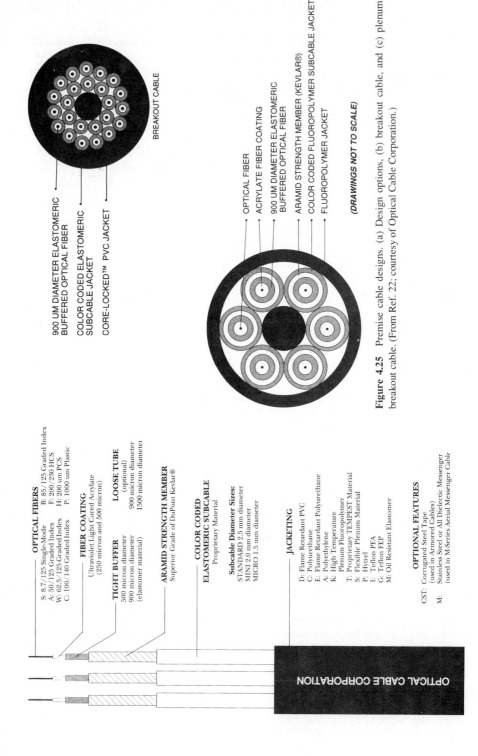

OPTICAL FIBERS

S: 8.7/125 Single-Mode   B: 85/125 Graded Index
A: 50/125 Graded Index   F: 200/230 HCS
W: 62.5/125 Graded Index   H: 200 um PCS
C: 100/140 Graded Index   P: 1000 um Plastic

FIBER COATING
Ultraviolet Light Cured Acrylate
(250 micron and 500 micron)

TIGHT BUFFER
500 micron diameter
900 micron diameter
(elastomer material)

LOOSE TUBE
(optional)
900 micron diameter
1500 micron diameter

ARAMID STRENGTH MEMBER
Superior Grade of DuPont Kevlar®

**COLOR CODED
ELASTOMERIC SUBCABLE**
Proprietary Material

Subcable Diameter Sizes:
STANDARD 2.5 mm diameter
MINI 2.0 mm diameter
MICRO 1.5 mm diameter

JACKETING

D: Flame Retardant PVC
C: Polyurethane
E: Flame Retardant Polyurethane
A: Polyethylene
K: High Temperature
    Plenum Fluoropolymer
T: Proprietary TEMPEST Material
S: Flexible Plenum Material
P: Hytrel
I: Teflon PFA
G: Teflon FEP
M: Oil Resistant Elastomer

OPTIONAL FEATURES

CST: Corrugated Steel Tape
    (used in Armored Cables)
M: Stainless Steel or All Dielectric Messenger
    (used in M-Series Aerial Messenger Cable

OPTICAL CABLE CORPORATION

900 UM DIAMETER ELASTOMERIC
BUFFERED OPTICAL FIBER

COLOR CODED ELASTOMERIC
SUBCABLE JACKET

CORE-LOCKED™ PVC JACKET

BREAKOUT CABLE

OPTICAL FIBER

ACRYLATE FIBER COATING

900 UM DIAMETER ELASTOMERIC
BUFFERED OPTICAL FIBER

ARAMID STRENGTH MEMBER (KEVLAR®)

COLOR CODED FLUOROPOLYMER SUBCABLE JACKET

FLUOROPOLYMER JACKET

(DRAWINGS NOT TO SCALE)

Figure 4.25 Premise cable designs. (a) Design options, (b) breakout cable, and (c) plenum breakout cable. (From Ref. 22; courtesy of Optical Cable Corporation.)

106

TABLE 4.3 PREMISE CABLE CHARACTERISTICS

B-series breakout cables

FIBER COUNT	STANDARD (2.5 mm Subcable)				MINI (2.0 mm Subcable)				MICRO (1.5 mm Subcable)			
	Dia. (mm)	Wt. (kg/km)	Tensile Load Rating (N)* Short Term	Long Term	Dia. (mm)	Wt. (kg/km)	Tensile Load Rating (N)* Short Term	Long Term	Dia. (mm)	Wt. (kg/km)	Tensile Load Rating (N)* Short Term	Long Term
2	7.0	50	1,200	500	6.0	38	800	200	5.0	27	800	200
4	8.0	65	2,000	800	7.0	50	1,600	400	6.0	37	1,600	400
6	9.5	82	3,000	1,200	8.0	64	2,400	600	6.5	39	2,400	600
8	11.0	111	4,000	1,700	9.5	83	3,200	800	7.5	52	3,200	800
10	13.0	152	5,000	1,900	10.5	102	4,000	1,000	8.5	67	4,000	1,000
12	12.5	148	6,000	2,500	10.0	96	4,800	1,200	8.0	64	4,800	1,200
18	14.5	186	8,000	3,500	12.0	121	6,000	1,500	9.5	85	6,000	1,500
24	17.5	237	10,000	3,800	14.0	178	7,200	1,800	11.0	108	7,200	1,800
30	19.5	285	12,000	5,000	15.5	203	8,400	2,100	12.5	132	8,400	2,100
36	20.5	305	14,000	6,000	16.5	208	9,600	2,400	13.0	137	9,600	2,400
48	23.5	393	18,000	7,500	19.5	288	12,000	3,000	14.5	178	12,000	3,000
60	25.0	426	22,000	8,800	20.5	305	14,400	3,600	15.5	194	14,400	3,600
72	27.5	516	26,000	11,000	22.5	366	16,800	4,200	17.5	223	16,800	4,200
84					24.5	433	19,200	4,800	19.0	254	19,200	4,800
96					26.5	506	21,600	5,400	20.5	296	21,600	5,400
108					27.0	509	24,000	6,000	21.0	301	24,000	6,000
120									22.5	346	26,400	6,600
132									23.5	377	28,800	7,200
144									24.5	410	31,200	7,800
156									26.0	463	33,600	8,400

*Installation loads in excess of 2,700 N (600 lbs.) are not recommended.   Part Number example: B12-125D-W3SB/1UC/900
SPECIFICATION AND CABLE ORDERING GUIDE FOR FURTHER DETAILS

Simplex and duplex cables

SPECIFICATIONS		PVC OFNR		S-TYPE FLEXIBLE OFNP		HIGH TEMP PLENUM OFNP		FLAME RETARDANT POLYURETHANE		TEMPEST	
		SIMPLEX A01-030D	DUPLEX A02-030D	SIMPLEX A01-030S	DUPLEX A02-030S	SIMPLEX A01-025K	DUPLEX A02-025K	SIMPLEX A01-030E	DUPLEX A02-030E	SIMPLEX A01-030T	DUPLEX A02-030T
Diameter	mm	3.0	3.0X6.5	3.0	3.0X6.5	2.5	2.5x5.5	3.0	3.0X6.5	3.0	3.0X6.5
Weight	kg/km	8.0	16.0	9.0	18.0	7.0	14.0	7.5	15.0	7.0	14.0
Tensile Load:											
Short Term	N	500	1000	500	1000	500	1000	500	1000	500	1000
Long Term	N	300	500	300	500	300	500	300	500	300	500
Minimum Bend Radius:											
Loaded	cm	5		5		5		5		5	
No Load	cm	3		3		3		3		3	
Crush Resistance	N/cm	750		500		750		1000		1000	
Impact Resistance	Cycles	1000		200		1000		1000		1000	
Flex Resistance	Cycles	7500		2000		5000		10,000		10,000	
Temperature:											
Operating	°C	-40 to +85		-20 to +65		-20 to +85		-55 to +85		-55 to +85	
Storage	°C	-55 to +85		-40 to +65		-40 to +85		-70 to +85		-70 to +85	

These specifications are subject to change without prior notification.

Source: Ref. 22, Courtesy of Optical Cable Corporation.

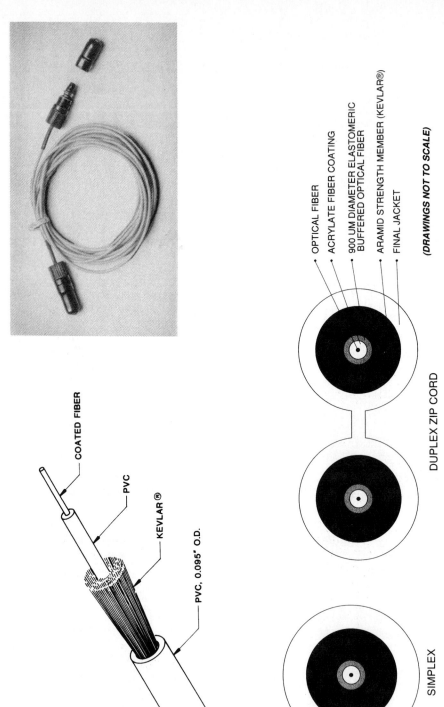

COATED FIBER

PVC

KEVLAR®

PVC, 0.095" O.D.

• OPTICAL FIBER

ACRYLATE FIBER COATING

900 UM DIAMETER ELASTOMERIC
BUFFERED OPTICAL FIBER

ARAMID STRENGTH MEMBER (KEVLAR®)

• FINAL JACKET

(DRAWINGS NOT TO SCALE)

DUPLEX ZIP CORD

SIMPLEX

Figure 4.26 Fiber jumper and "pigtail" cable design. (a) Single-fiber jumper cord, (b) single-fiber

ened and overjacketed with heavy color-coded buffers can be broken out and handled individually.

4.5.5.2 Building Cable

Building cable is generally medium strength cable that can be pulled horizontally and vertically through plenum or open ceiling and floor areas. Being in open areas it is therefore required to have a fire-retardant outer jacket. Building cable is generally not gel filled.

Building cable is generally of the tight buffered core design and contains color-coded individually buffered fibers so that the fibers can be fanned out of the cable and individually connectorized.

The fiber is generally a multimode design compatible with LANs or relatively low-speed terminal equipment, although it can be single mode if the application requires it. Fiber count varies by manufacturer but generally is available with 2 to more than 150 fibers.

4.5.5.3 Pigtails and Jumper Cords

Pigtails and jumper cords are used to connect communications equipment or terminate cable to communications equipment at relatively short distances.

These cords or pigtails contain generally one or two fibers, although some designs contain four. They are heavily buffered with fire-retardant PVC (or other materials as required by the National Electric Code) and often contain a dielectric strength member (such as Kevlar® so that they can be handled, routed through equipment bays, and terminated within splice enclosures and optical connector bays.

Typical designs are illustrated in Fig. 4.26. Fiber can be jacketed so as to be pulled apart like "zip" cord or with an overjacket that can be removed with a rip cord.

Pigtails and jumpers have a connector on one or both ends and a bare fiber on the other. Pigtails are spliced to entry or riser cables to terminate these cables. Jumpers have a connector on both ends. Jumpers are used for optical connections between connectors on an optical patch panel or between optical patch panels and terminal equipment.

Both are of a tight buffer design often overstranded with Kevlar strengthening materials, and overjacketed with a PVC or polyurethane jacket.

REFERENCES

1. Fujitsu product literature, n.d.

2. Les Borsuk, "Introduction to Fiber Optics—What All Connector Engineers Need to Know," ITT Cannon Electric, Santa Ana, Calif., n.d.

3. CCITT Recommendation G651.

4. ITT product sheets, 1978.

5. Lee L. Blyler, Jr., and Shiro Matsuoka, "Polymer Protection for Glass Fibers," *Bell Laboratories RECORD,* Dec. 1979, pp. 315–319.

6. "Application Notes, Introduction to Fiber Optics," T&B/Ansley Publication No. AFO-1000, 1979. (Drawings reproduced with permission from Thomas & Betts Corp.)

7. "Introduction to Fiber Optics and AMP Fiber-Optic Products," HB 5444, AMP Incorporated, n.d.

8. Charles Kleekamp and Bruce Metcalf, "Designer's Guide to Fiber Optics—Part 1," *EDN,* Jan. 5, 1978.

9. Herb Lubars, "Optical Fiber Cable Systems," presented at the Western Regional Meeting, AAR, May 1, 1979.

10. R. Hoss, *Fiber Optic Communications Design Handbook* (Englewood Cliffs, N.J.: Prentice Hall, Inc., 1990), pp. 113, 154, and 106.

11. M. K. Barnoski, *Fundamentals of Optical Fiber Communications* (New York: Academic Press, 1976).

12. D. Gloge, *Bell System Technological Journal* Vol. 54, 1975, p. 243.

13. W. B. Gardner, *Bell System Technological Journal* Vol. 54, 1975, p. 457.

14. R. Hoss, *Fiber Optic Communications Design Handbook,* (Englewood Cliffs, N.J.: Prentice Hall, 1990), p. 12.

15. G. S. Anderson, J. C. McNaughton, and R. L. Ohihaber, "Fiber Optic Cables for Telecommunications," Third International Telecommunication Exposition, Dallas, Tex., 1979.

16. Northern Telecom Lite Star Optical Fiber Cable, sales brochure 14050/05, issue 2.

17. C. K. Kao, "Optical Fiber Communications Technology," *Electrical Communication,* Vol. 54, No. 3, 1979.

18. Peter R. Bark, Ulrich Oestreich, and Gunter Zeidler, "Fiber Optic Cable Design, Testing and Installation Experiences," Siecor Optical Cables, Inc., Horseheads, NY, and Siemens AE, Munich, W. Germany, Nov. 1978.

19. C. W. Kleekamp and B. D. Metcalf, "Fiber Optics for Tactical Communications," Air Force Electronic Systems Division Report ESD-TR-79-121, Mitre Corp., Bedford, Mass., MTR-3723, Apr. 1979, p. 34.

20. Drawing based on Northern Telecom product, provided courtesy Northern Telecom.

21. AT&T Lightpack product literature, 1985.

22. Optical Cable Corporation product literature, P.O. Box 11967, Roanoke, VA 24022-1967.

23. AT&T Product Application Bulletin 626-108-110, "Single-Mode Lightguide Cable," May 1986.

5

Splices, Connectors, and Couplers

In fiber optic systems major light losses can occur at three optical junctions.

1. From source to fiber
2. From fiber to fiber
3. From fiber to photodetector

Whether these junctions are permanent splices or demountable connectors, considerable ingenuity has been used in the design of joining devices to keep the losses at a minimum.

Junction losses are most often the result of combinations of both intrinsic and extrinsic loss mechanisms, defined as follows:

1. *Intrinsic.* Tolerance mismatches intrinsic to the fiber such as core size mismatch, NA mismatch, core concentricity offset, and graded-index profile differences
2. *Extrinsic.* Loss mechanisms external to the fiber and associated with fiber end preparation and connector or splice tolerance variations such as lateral offset, angular misalignment, fiber end separation, reflection, and surface roughness

Figure 5.1 illustrates these mechanisms, and Figs. 5.2 and 5.3 illustrate the magnitude of the insertion loss created by each individually [1, 2].

When connectors or splices are joined, these mechanisms interact in somewhat of a random fashion. For that reason connector and splice loss is generally character-

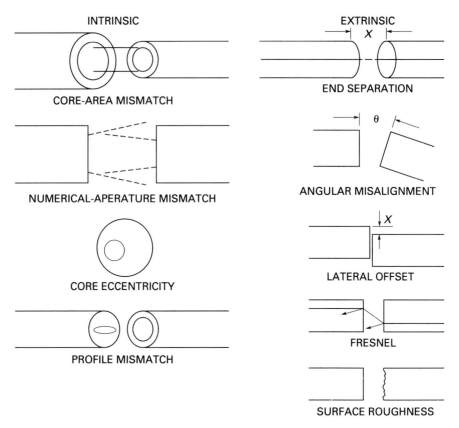

Figure 5.1   Coupling-loss mechanisms. (From Ref. 1; copyright 1990 Prentice Hall, Inc.)

ized not as an absolute value, but as a range of values. This range is called a histogram, plotted as attenuation versus number of occurrences. Occurrences are either the number of matings of the same or different connectors, or the number of splices with the same fiber. The resultant histogram can be related to a statistically "normal distribution." This distribution will be explained in Chapter 7 (see Fig. 7.6). In doing so the splice or connector loss can be described in terms of a mean loss (average or typical loss) with the maximum losses represented as an expected deviation from the mean. The term *standard deviation* is used to describe the expected deviations.

As can be seen in Fig. 7.6, the mean relates to the attenuation value that 50% of the joints would achieve. One standard deviation (SD) relates to the attenuation value that 84.1% of the joints will fall within. The maximum loss value specified for a joint is most commonly 3 SDs from the mean. This is the value that 99.87% of the joints fall within.

Figure 5.4 illustrates the sort of distribution that may be observed for common connectors and splices. It is not always the shape of a "normal" distribution because it may be common to have several loss occurrences at or close to zero.

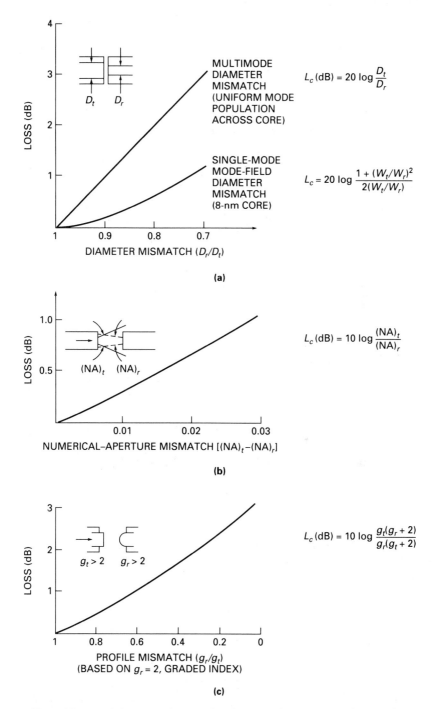

Figure 5.2 Intrinsic loss mechanisms. (a) Coupling loss owing to core-area, (D) or mode-field diameter (W) mismatch, (b) coupling loss as a function of NA mismatch, and (c) coupling loss attributed to profile mismatch. (From Ref 1; copyright 1990 Prentice Hall, Inc., pp. 122–124). Drawings originally provided by Dennis Knecht.

113

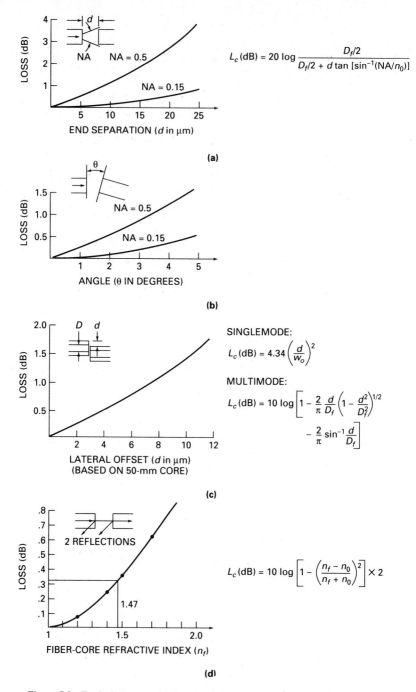

Figure 5.3 Extrinsic-loss mechanisms. (a) Coupling loss as a function of end separation, (b) coupling loss as a function of angular misalignment, (c) coupling loss to lateral offset or core/clad concentricity, and (d) coupling loss owing to Fresnel reflection. (From Ref. 1; copyright 1990 Prentice Hall, Inc., pp. 123–126). Drawings originally provided by Dennis Knecht.

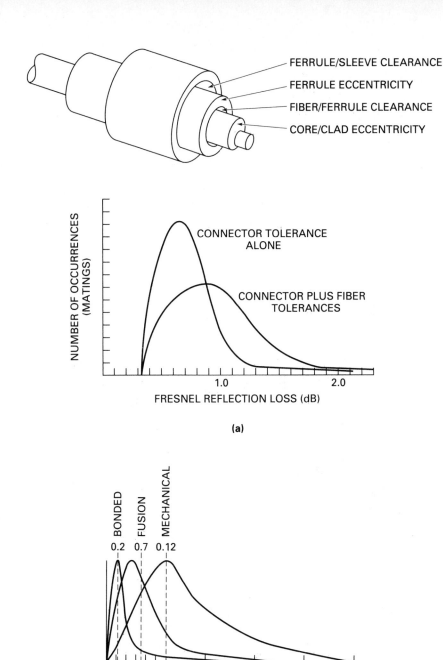

Figure 5.4 Insertion loss histograms for typical connectors and splices. (a) Connector loss mechanisms showing the effect of connector tolerances alone as well as connector plus fiber tolerances and (b) splice loss characteristics typical of various splicing technologies. (From Ref. 1, copyright Prentice Hall, Inc., pp. 127, 133.)

Note also from Fig. 5.4 that the losses measured for the connector and splice tolerances alone (the values given on most supplier data sheets) are somewhat less than those actually observed in practice. This is because nearly all suppliers measure joining loss using the same fiber, broken and rejoined. In this way most of the intrinsic fiber losses are eliminated from the measurement. In practice, fiber from different batches are joined and therefore the intrinsic losses will add to the losses advertised by the supplier. In the case of a splice, if the alignment is active, the added loss will be minor. In the case of a connector it is not possible to know the total effect of the fiber tolerances. A good conservative measure is to increase connector only loss by a factor of 1.5 to 2 on the transmit end and where fiber-to-fiber connections are made. Generally receivers use a larger-core fiber into the detector and thus connector loss is minimized.

Note that just as joining loss mechanisms combine in a random statistical manner, losses from individual joinings are represented statistically when added in series. When calculating the total loss that would result from two or more connectors or splices they are summed in the following fashion. Mean values are added together directly. Standard deviations are added as the square root of the sum of the squares.

$$\text{Total mean loss} = \text{number of joints} \times \text{mean loss per joint}$$

$$\text{Total SD} = \sqrt{\text{number of joints}} \times \text{SD per joint}$$

Maximum loss is generally defined as 3 SDs (representing 99.87% of the joints) or

$$\text{Max loss} = \text{total mean loss} + 3 \times \text{total SD}$$

5.1 SPLICING OPTICAL FIBERS

5.1.1 Fusion Splicing

A fusion splicer [3] (Fig. 5.5) is a portable self-contained unit that aligns two optical fibers and fuses (melts) them together with an electric arc. It generally contains the following equipment and features:

1. Fiber sheath removing tools and buffer removing chemicals (such as methylene chloride). Typically the buffered fiber is pushed into a small hole of a container with a stripping fluid, left for about 10 s and removed. After about 1 min, the softened buffer is then removed with a wipe containing a cleaning fluid (generally reagent-grade isopropyl alcohol).

2. Precision fiber end cleaver to prepare the fiber end with a clean break and a smooth end surface perpendicular to the fiber axis by less than 1 degree. This cleaver is often built into the splice machine to reduce the amount of fiber handling.

3. Fiber holder and alignment fixture. Generally these are two V-shaped grooves, one for each fiber, with the fiber held in place by the friction of a clamp (often a rubber magnet).

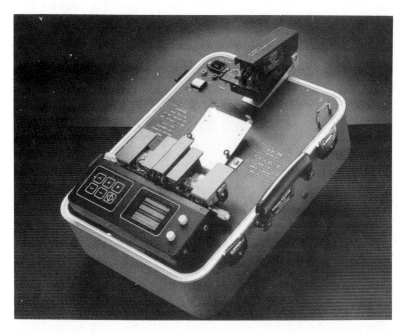

Figure 5.5   Fusion splicer. (From Ref. 3; reprinted courtesy of Siecor.)

4. Microscope or projection image screen to view and position the fibers. Generally the fiber is placed in the holding fixture and moved manually to a position marked by stationary reference lines in the viewing screen.

5. A means for injecting light into one of the fibers, and locally detecting either coupled light transmitted to the other fiber, or noncoupled light reflected off the end surface. The detector measurement is used by the automatic alignment mechanism of the splicer to determine the optimum alignment point.

6. An automatic alignment mechanism within the splicer that, on pushing a button, aligns the two fibers at the point where the most light is transmitted from one core to the other. This is done in some machines by coupling light at the transmit end of the span and locally detecting the light reflected off the end of that transmit fiber. The detector is within the fiber-holding fixture on the splicer. When the transmit fiber is properly aligned with a mating fiber, the light will couple through to the other core, and the reflection will reduce to a minimum.

7. An automatic fusing mechanism that performs the following: brings the fibers to within a fixed distance of each other; arcs across the end surfaces to burn off small hackles and dirt; moves the fibers together with a given pressure; and arcs across the joint with a fixed time and temperature (current) to fuse the fibers precisely. Fusion current and time is generally precision adjustable for each fiber product and type, as well as to compensate for environmental conditions. Figure 5.6 illustrates the fusion steps normally involved with an automated fusion splicer.

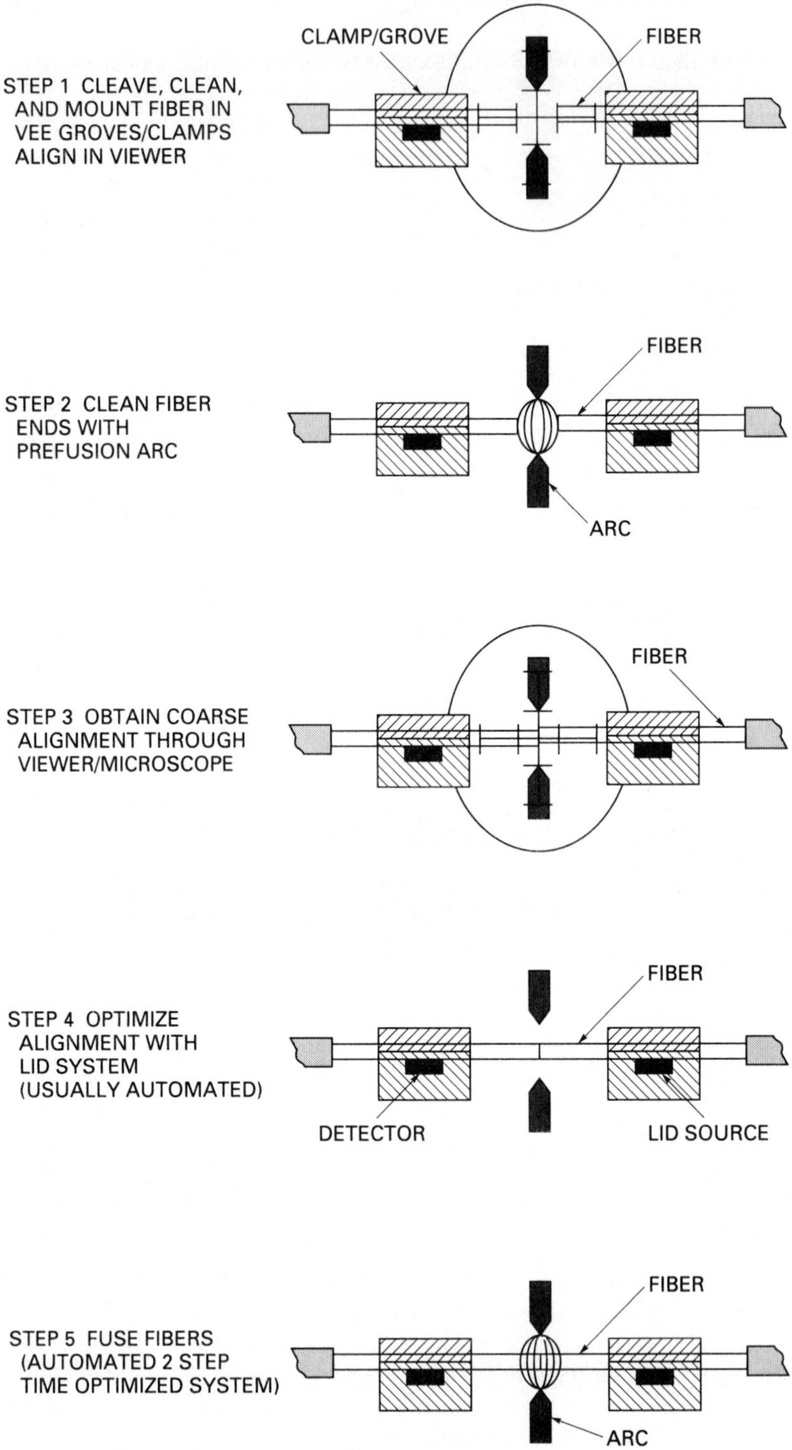

STEP 1 CLEAVE, CLEAN, AND MOUNT FIBER IN VEE GROVES/CLAMPS ALIGN IN VIEWER

CLAMP/GROVE FIBER

STEP 2 CLEAN FIBER ENDS WITH PREFUSION ARC

FIBER

ARC

STEP 3 OBTAIN COARSE ALIGNMENT THROUGH VIEWER/MICROSCOPE

FIBER

STEP 4 OPTIMIZE ALIGNMENT WITH LID SYSTEM (USUALLY AUTOMATED)

FIBER

DETECTOR LID SOURCE

STEP 5 FUSE FIBERS (AUTOMATED 2 STEP TIME OPTIMIZED SYSTEM)

FIBER

ARC

Figure 5.6 Fusion splicing process.

8. A spliced fiber overcoating and protection kit. Once the fiber is spliced the bare fiber must be removed and overcoated with a substance which protects it from dirt and abrasion. The bare fiber may also be placed inside a rigid sleeve, sandwich-style fixture or holding assembly that further protects the fiber from mishandling and foreign materials. This assembly is often a part of the splice tray that fits within the splice enclosure. The enclosure is described in Chapter 8. A splice tray and organizer [29] is illustrated in Fig. 5.7. Fiber buffer tubes that protect the fiber when stowed within the enclosure are placed over each fiber end in the end preparation process before splicing.

The specifications for the Siecor fusion splicer with active alignment (LID) are given in Table 5.1 [3].

The Siecor Model M90 Microprocessor Controlled Fusion Splicer incorporates both a Profile Alignment System (PAS) and LID-System unit in one machine. These systems together allow the M90 to use the PAS to splice pigtails with 900-μm coatings or hermetic coatings, while offering the 1300-nm LID-System unit for optimizing and measuring splices [3].

Also included is a Fusion Time Optimization Program, whereby the LID-System unit in the M90 monitors the power through the splice during fusion. At the point of minimum splice loss, a microprocessor shuts off the fusion arc to achieve the optimal splice given the conditions present.

According to Siecor, average splice loss using the M90 on identical single-mode fibers is 0.03 dB.

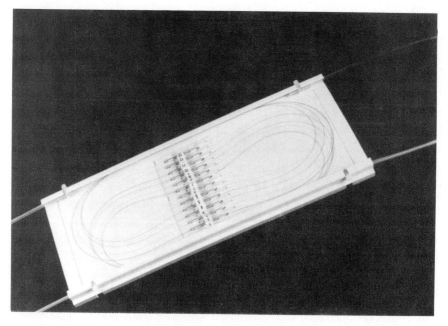

Figure 5.7   Splice tray and organizer. (From Ref. 29; courtesy of Northern Telecom.)

TABLE 5.1 TYPICAL PERFORMANCE SPECIFICATIONS FOR SIECOR MODEL M90 FUSION SPLICE EQUIPMENT

Variable parameters		
Parameter adjustment	Range of adjustment	Step size
Cleaning time	0.05–0.3 s	0.01 s
Cleaning current	10.0–61.0 mA	0.1 mA
Fusion time	0 –10.0 s	0.1 s
Fusion current	10–20 mA	0.1 mA
Axial distance apart before fusion	2–10 μm	0.1 μm
Automatic feed-forward	0–10 μm	0.1 μm
Prefusion time (time between start of fusion and automatic feed-forward)	0.16–2.50 s	0.01 s
Prefusion current	10–20 mA	0.1 mA

Specifications	
RAM	2 Kbytes expandable up to 32 Kbytes with battery back-up
ROM	Up to 64 Kbytes
Fiber clamps	Fiber diameters 125 μm to 900 μm (bare fiber or primary coating) using two sets of interchangeable clamps
Fiber positioning	Three axes; high-resolution piezoelectric adjusting elements: resolution better than 0.1 μm; adjustment range: X and Y axis more than 120 μm; Z-axis more than 40 μm
High-contrast LCD display	High-contrast image, magnification 60X, illuminated, X and Y viewing axes displayed simultaneously, two-line message display
Circuitry	Modular, plug-in architecture with 2 vacant expansion slots for future upgrades
Operating temperature	−5°C to +45°C
Power supply	AC: 110V to 130V/200V to 240V AC; automatic voltage range selection; 50 to 60 Hz; 70 W max.
	DC: Rechargeable 12V, 6.7 Ah battery, about 100 field splices per battery charge; charging time 3 hours for fully discharged battery with built-in charging unit; deep-discharging protection; separate input for external 12V DC supply, polarity protected
Arc	High-frequency AC voltage; pre-fusion and fusion current adjustable, 10 to 20 mA
Dimensions	42 × 32 × 18 cm (16.5" × 12.6" × 7.0")
Weight	16 kg (35 lbs)

Integrated LID-SYSTEM® Unit
LED wavelength 1300 nm; will function with any single-mode or multimode fiber; external diameter (with primary coating) 250 μm to 500 μm; refractive index of primary coating greater than that of the cladding glass.

Source: Ref. 3; courtesy of Siecor.

*LID-SYSTEM® is a registered trademark of Siecor Corporation. Siecor reserves the right to improve, enhance, and modify the features and specifications of Siecor products without prior notification.

5.1.2 Mechanical Splice

Mechanical splices are typically defined as any type that is not fusion, where splicing is usually accomplished by inserting the two prepared fiber ends into a fiber alignment mechanism and fixing or gluing them in place. A category of mechanical splice, called a "rotary splice" will be discussed in the next section. The type of mechanical splice described in this section aligns to the fiber outer surfaces; thus the insertion loss is subject to both fiber concentricity tolerances as well as alignment sleeve tolerances.

Although the stability and performance of mechanical splices is quite good, it is generally less than that of fusion, and therefore used when a few splices are to be done where the cost, setup, or training required of the fusion splice equipment is not justifiable.

Some mechanical splices use ultraviolet-cured epoxy to hold the fibers within the alignment mechanism or sleeve. In this case the housing is glass to transmit the ultraviolet light from the curing lamp. Others use heat-cured epoxy and yet others a clamping or friction mechanism to hold the fibers in place. The splice kit generally contains the following:

1. A set of splice sleeves, perhaps with bonding agent inside. The sleeves align the fibers, hold them in place via epoxy or friction, and have some friction or crimp-type mechanism for securing themselves to the fiber buffer material as well. Different housings are sold for different fiber buffering, outer diameter glass, and core size/type (single or multimode).

2. A fiber coating stripper tool and/or stripping chemicals.

3. A fiber-end cleaving tool.

4. Hand-held microscope and lamp to inspect fiber-end finish after cleaving.

5. A fiber splice assembly fixture to hold the housing and fibers in place during the splicing operation.

6. An ultraviolet curing lamp, or a heating oven in the case of epoxy type assemblies.

7. Adhesive or epoxy, if applicable.

8. Possibly a crimping tool to crimp the splice assembly to the fiber jacket where individually jacketed fibers are used.

The technologies used to align the fibers within the splice sleeves are the key to the performance of the splice.

In GTE's Elastomeric splice [5] when fibers are inserted into the Elastomeric sleeve, the sleeve aligns them by exerting equal pressure from three elastomeric elements towards a central axis. The fibers are epoxied once positioned.

The 3M Fiberlok mechanical splice, illustrated in Fig. 5.8(a), also aligns the fibers using three points of pressure toward a central axis [26]. It splices both multi-

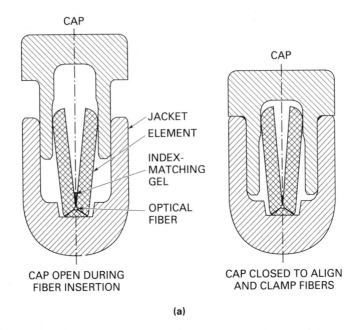

CAP

CAP

JACKET
ELEMENT

INDEX-
MATCHING
GEL

OPTICAL
FIBER

CAP OPEN DURING
FIBER INSERTION

CAP CLOSED TO ALIGN
AND CLAMP FIBERS

(a)

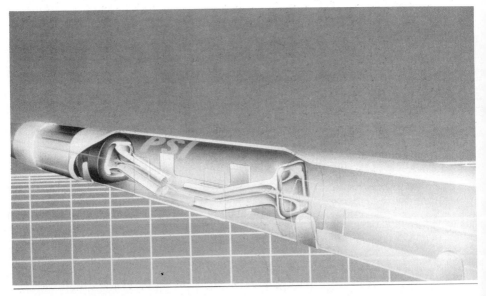

Figure 5.8 Fiber mechanical splice alignment techniques. (a) 3M Fibrlok and (b) PSI lightlinker fiber optic splice system.

mode and single-mode fiber. The fibers are cleaved, cleaned, and set into the plastic alignment assembly and surrounded by index-matching gel. When the cap is pushed on, the plastic assembly presses onto both fibers at the three points as shown in the diagram. Friction holds them in place with tensile loads of from 2 to 5 lb.

The splice marketed by PSI [4] uses a patented four-rod glass alignment guide. As Fig. 5.8(b) illustrates, the fiber ends are guided into an opening that narrows down into a channel formed by four glass rods. The channel takes a bend forcing the fiber to center itself in a V-shaped groove formed by two of the rods. The opposing fiber is channeled into the same V-shaped groove, and thus the two align with each other, two similar points on the surface of each fiber being the alignment reference.

Typical mechanical splice performance is as follows:

1. *Loss.* For the elastomeric splice, single-mode fibers exhibited mean losses in the range of 0.2 dB with maximums of about 0.5 dB [2] when the fibers were rotated to achieve optimum throughput. The 3M splice is advertised to achieve losses less than 0.2 dB, but losses of 0.01 to 0.07 dB are typical [27]. Tests of single-mode splices made with dissimilar fibers from dissimilar manufacturers resulted in mean losses of 0.15 dB and a range of from 0 to slightly greater than 0.3 dB [27]. For multimode fiber, mean losses of 0.08 to 0.2 dB, and maximums between 0.3 to 0.6 dB are achievable depending on the product, the fiber size, and the tuning [2, 4].

2. *Temperature effects.* Temperature cycling between −55°C and + 85°C induced changes on the average of about 0.05 dB with an absolute range of around ± 0.3 dB with single-mode fiber [2]. Temperature cycling between 25°C and 80°C on the 3M splice resulted in variations of +0.11 and −0.12 dB [27].

3. *Long-term effects.* There was no significant change in transmission with their design after a 4-year burn-in cycle (−50°C to +70°C) and 3-year storage period [2].

5.1.3 Rotary Splice

Rotary splicing is a unique mechanical splicing system, developed by AT&T [6] that works on a different principle than the design discussed earlier. With this design the fibers are mounted in glass holders, the ends polished much like a connector, the two cores actively aligned, and the entire assembly bonded together with epoxy. Because this approach employs active alignment, it has a very low loss.

The ferrule assemblies and alignment sleeve are shown in Fig. 5.9. The completed assembly is stowed within a special AT&T splice tray that keeps a spring-loaded compression on the assembly. The tray is similar in appearance to the one in Fig. 5.6.

A summary of the splicing procedure and equipment used is as follows [6]:

1. A buffer tube is placed over the fiber ends, and the ends are prepared by removing the coating as with fusion splicing.

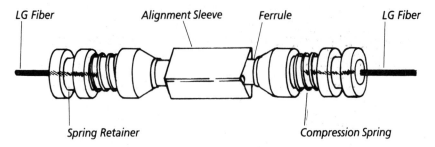

LG Fiber   Alignment Sleeve   Ferrule   LG Fiber

Spring Retainer   Compression Spring

Figure 5.9   Rotary mechanical splice. (Reprinted Courtesy of AT&T.)

2. A glass ferrule is filled with ultraviolet-cured epoxy with a syringe and the fiber inserted into it so that the fiber sticks out the mating end. The epoxy is then cured using an ultraviolet flash lamp that comes with the splice kit.

3. The fiber-end sticking out of the ferrule is then scored with a cleaving tool, pulled and removed, leaving a smooth fiber surface just outside the mating end of the ferrule.

4. The ferrule assembly and fiber end are then manually polished in a two-stage process using wetted 8- and 1-μm paper.

5. The polished ferrules with fiber attached are then inserted into the alignment sleeve with the aid of a tool that holds and separates the sleeve, and index-matching gel (which is also a heat-cured epoxy) is placed between the ends of the ferrules.

6. The transmit fiber is illuminated, the alignment test set is calibrated, and the receive end of the ferrule and alignment sleeve assembly is inserted into a housing that holds the detector. A tuning adaptor (a fixture to hold on to) is placed over the transmit ferrule.

7. The ferrules are made with a slight eccentricity so that when the tuning adapter rotates the transmit ferrule, there will be a point at which a minimum reading occurs at the detector. This is the point at which the most light is coupled through to the receiving fiber core, and the least light is scattered into the detector. After a four-step progressive tuning process the test set reads the actual loss. When it reaches less than 0.07 dB the process stops. The AT&T procedure calls for four attempts with different alignment sleeves until this loss level is reached.

8. The splice is then removed from the test set and tuning adapter, and placed in a specially designed splice tray, using a compression tool to handle it and insert it into the tray. Springs keep the completed splice compressed together.

9. Once all splices are in the tray, the assembly is placed into a heat-curing oven for 15 min and the index-matching gel epoxy cures to hold the ferrules together.

5.2 CONNECTORS

Connectors are used whenever two fibers, or a fiber and an electro-optical source or detector, are to be joined and disconnected repeatedly. This is generally at fiber terminal equipment, optical patch panels, or fiber couplers within a LAN.

Each connector in a fiber-optic system introduces an optical power loss. Connectors are present at the transmitter and receiver interface as a minimum. If jumpers and an optical patch panel are used to connect the optical cable and the equipment, then the number of connectors on each end can double. For this reason the proper choice of connectors can make a significant difference in total system performance margins. Connector losses typically range from 0.1 to 0.7 dB based on measurements with the same fiber cut and reconnected. The values given in this section are for connector loss measured in this fashion. With fibers from different lots, and with wear, the loss can increase beyond these levels, perhaps double. In terms of percentage power loss,

$$0.05 \text{ dB} = 1\% \text{ loss}$$

$$0.09 \text{ dB} = 2\% \text{ loss}$$

$$0.23 \text{ dB} = 5\% \text{ loss}$$

$$0.50 \text{ dB} = 10\% \text{ loss}$$

$$1.00 \text{ dB} = 20\% \text{ loss}$$

The challenge in designing a connector is to make (1) a ferrule with a small, precision fiber-accepting hole that matches the fiber outer diameter, and that is concentric with the outer mating surface of the ferrule; (2) a mating structure that maintains offset to within a few percent of core diameter and angular alignment within less than a degree from mating to mating, but that is not so tight fitting that friction and abrasion cause metal flaking and wear; and (3) a mating surface arrangement that ideally leaves no reflective air gap between fibers but protects fiber ends from the scratching that can occur if fibers touch. It is also desirable that the design is a recognized standard. This ensures compatibility between manufacturers. Some of the more common designs that have addressed these problems will be discussed later. These include the low-cost resilient ferrule design, the SMA design, the biconic, the ST type, the SC type, and the FDDI connector.

For trunking systems, generally single-unit connectors are appropriate. For premise cabling, the ferrule and alignment sleeves are housed in a duplex assembly. In addition to the FDDI standard, much of the international ISO/IEC* standardization activity, at the time of this printing, is focusing on the ST and SC connectors for premise cabling.

*ISO = International Standards Organization; IEC = International Electro-Technical Commission.

5.2.1 Resilient Ferrule Design

A design that has been for many years a very popular and low-cost connector for large-core multimode and plastic fiber is a plastic "resilient ferrule" connector that mates in a tapered sleeve (illustrated in Fig. 5.10). Information from Amp [8–10] is used to describe this connector.

The connector uses a flexible plastic material for the ferrule, and metal for the construction of the threaded cap and retaining assembly. It also includes a fiber strain

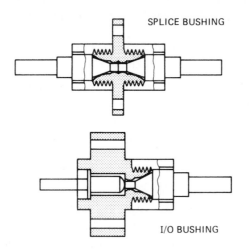

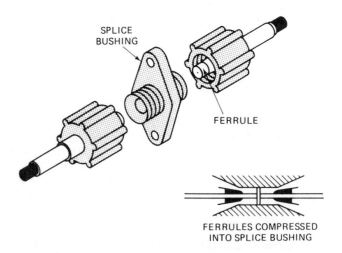

Figure 5.10   Resilient ferrule alignment mechanism. (From Ref. 8; courtesy of AMP.)

relief that holds the cable strength members with a mechanical crimp applied to the metal eyelet.

Fiber alignment is repeatable because of the tight interference fit of the ferrule into the splice housing. Tolerance mismatch of the fiber diameter is absorbed by the resiliency of the plastic elastomeric ferrule. Tolerances on the parts are not critical, therefore permitting them to be manufactured economically. The only ultraprecise operation required is the placement of the hole in the ferrule in relationship with the true center that exists when the connector is mated (a proprietary AMP manufacturing technique). Essentially the plastic ferrules are compressed into a circular drill fixture that forces them to assume their in-service shape. The fiber-receiving hole is then drilled at the dead center of the blank. The technique ensures that the hole is on true center regardless of the size of the fiber or initial shape of the ferrule.

The connector contains the following parts:

1. *Ferrule.* Holds the fiber and serves as the resilient part of the alignment mechanism.
2. *Cap.* Provides the means of maintaining the connection.
3. *Crimp ring.* Provides the pressure to hold the fiber in the ferrule.
4. *Polishing bushing.* Used to polish the fiber to the correct length and finish; it is discarded after use.

The connector is mated either to a splice bushing for fiber-to-fiber connections or to an input/output (I/O) bushing containing an active device. It may also be mated to the active-device connector described later.

The ferrule is the resilient alignment mechanism. The ferrule and bushing achieve the alignment. The inside of the bushing is tapered, and as the cap is screwed on, the taper compresses the ferrule. This compression moves the fiber on center. When both connectors are joined in the bushing, the fibers are aligned, as shown in Fig. 5.10. The compression of the ferrule seals the fiber-to-fiber interface by providing a tight fit between the ferrule and fiber, and between the ferrule and bushing. The hole in the ferrule is typically slightly larger than the fiber, because the ferrule will be compressed tightly around the fiber. Differences in fiber sizes are thus compensated for.

Ferrules are available for a variety of fiber sizes. They are color-coded according to their hole size. Except for the ferrule, the parts of the single-position connector are common to all. A typical termination, shown in Fig. 5.11, begins by stripping a half-inch of the fiber's jacket. If required (by the fiber manufacturer), epoxy is applied to the fiber. The fiber is slid into the assembled connector and protrudes from the front of the ferrule. The crimp ring is crimped to the ferrule. The ring has two purposes: providing mechanical pressure to hold the fiber in the ferrule and retaining the cap. After crimping, the polishing bushing is screwed onto the cap. The epoxy is allowed to set, and the fiber and ferrule are polished flush with the bushing. After polishing, the bushing is removed and discarded. The connector is ready for use.

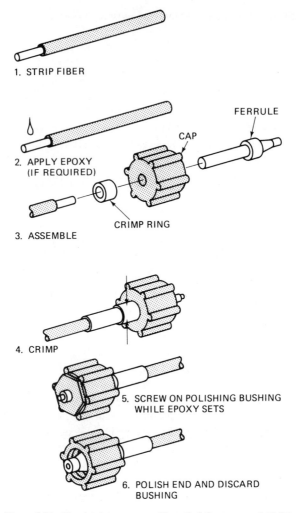

1. STRIP FIBER

2. APPLY EPOXY
   (IF REQUIRED)

3. ASSEMBLE

4. CRIMP

5. SCREW ON POLISHING BUSHING
   WHILE EPOXY SETS

6. POLISH END AND DISCARD
   BUSHING

Figure 5.11   Termination sequence. (From Ref. 8; courtesy of AMP.)

Splice bushings are either free hanging or bulkhead mounted. I/O bushings, although generally similar, are available in a variety of styles to fit different needs. Some have different cavity dimensions to accommodate different TO cans; one style fits a molded-lens package. Another style uses a press fit to hold the active device firmly. The I/O bushings are mounted over the active device and secured to the printed-circuit board.

AMP's small-fiber connector terminates 125- to 245-μm fibers. Like the single-position connector, the small-fiber connector uses a resilient ferrule to achieve alignment. To ensure the alignment of such small fibers, AMP has developed a proprietary method of placing the hole in the ferrule so that the fiber is centered when it is engaged.

The connector consists of the following parts:

1. *Cap*. Provides the means of maintaining the connection.
2. *Retaining sleeve*. Provides the pressure to hold the fiber in the ferrule; comes attached to the cap.
3. *Ferrule*. Holds the fiber and serves as the resilient part of the alignment mechanism.
4. *Heat-shrink tubing*. Shrunk around the retaining sleeve and fiber jacket to give extra support and strain relief.
5. *Polishing bushing*. Used to polish the fiber to the correct length and finish.

In a typical termination (Fig. 5.12) the fiber's strength members are placed over the rear end of the ferrule. The retaining assembly (cap and sleeve) is slid over the strength member, and the sleeve is crimped to the strength members. The shrink tubing is then applied. The tubing gives additional reinforcement for holding the fiber in place and preventing sharp bending at the rear of the sleeve.

The hex-shaped cap shown in Fig. 5.12 is metal to provide shielding of the source and detector. Both the ferrule and the metal retaining sleeve vary according to the fiber used. Ferrules are color coded.

Although the small-fiber connector is intended to be used with the active-device connector, it is also compatible with the bushings used with the single-position connector.

The Optimate active-device connector, shown in Fig. 5.13, is an improved design for connecting a ferrule active device to the small-fiber or single-position connector. The active device is held in the connector by a press-on retention plate. The inside of the connector is tapered at both ends. Because the semiconductor is connected to the end of the ferrule by a fiber, the final alignment is an efficient one of fiber to fiber.

An AMP-FIT hand tool with the CERTI-CRIMP ratchet is used to crimp Optimate ferrules, except for small-fiber ferrules. The tool has two crimping areas: one for single-position and multiple-position ferrules and one for Optimate Multimate ferrules. A CHA-MP hand tool is also available for all these ferrules. The small-fiber connector is crimped with an AMP commercial hand tool with removable die sets.

Note that these single-fiber connectors require the fibers to be epoxied in position. After the epoxy has set, the fiber end must be polished square and smooth. In some cases a 50 × microscope is necessary for checking the quality of the polish. The fiber end must be flush with the connector butt. For minimum losses the fiber ends should touch when the connectors are joined; however, in practice a gap must be maintained between the two surfaces, as repeated matings and vibrations will chip and scratch the surfaces. Any such damage to the surfaces would cause coupling loss. A typical gap distance is 0.0005 in.

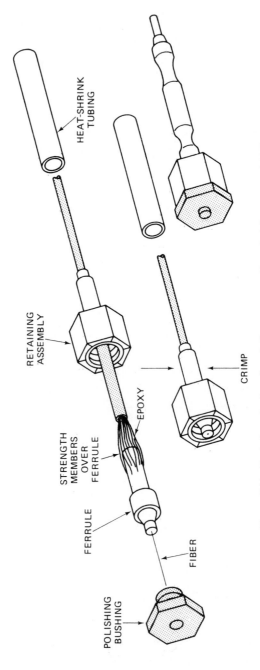

Figure 5.12 Typical termination of small-fiber connector. (From Ref. 8; courtesy of AMP.)

HEAT-SHRINK TUBING

RETAINING ASSEMBLY

STRENGTH MEMBERS OVER FERRULE

EPOXY

FERRULE

FIBER

POLISHING BUSHING

CRIMP

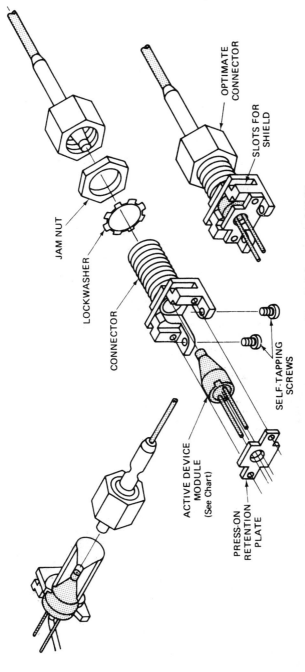

JAM NUT

LOCKWASHER

CONNECTOR

OPTIMATE
CONNECTOR

SLOTS FOR
SHIELD

SELF-TAPPING
SCREWS

ACTIVE DEVICE
MODULE
(See Chart)

PRESS-ON
RETENTION
PLATE

Figure 5.13 Active-device connector. (From Ref. 8; courtesy of AMP.)

5.2.2 SMA Design

The SMA connector design (illustrated in Fig. 5.14) is an adaption of the hex-nut connector used originally with high-bandwidth electronics, and it rapidly gained acceptance as a multimode fiber connector for most early fiber systems. Until about 1988, when it began to be surpassed by the ST, it enjoyed the position as the design with the greatest installed base.

Alignment in most SMA designs simply depended on a precision ferrule mating with tight tolerances into a precision sleeve. Interesting ferrule designs emerged to place a precision concentric hole in the center of the relatively large ferrule. Be-

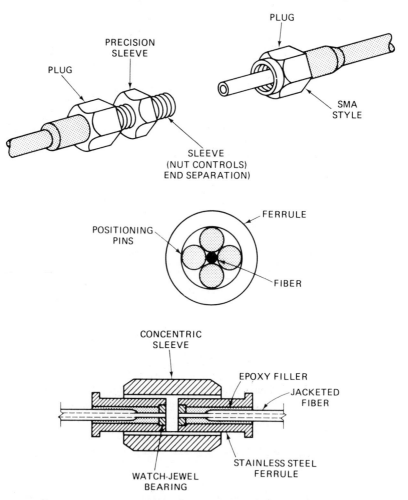

Figure 5.14 SMA connector design. (a) Straight-sleeve connector, (b) four-pin method connector (From Ref. 1 [a, b]; courtesy of AMP); and (c) watch-jewel–bearing connector designed by ITT Cannon. (From Ref. 23; copyright 1978 Milton S. Kiver Publications, Inc.)

cause it is easier to drill or manufacture a large concentric hole than a small one, most manufacturers would start with a large hole and insert various mechanisms in it for concentric fiber acceptance. ITT Cannon uses watch jewels of various sizes. Thomas & Betts uses a glass capillary. Amphenol's most popular design, the 906, uses four precision pins of various diameters to create concentric fiber acceptance holes of varying diameter.

Insertion loss with 62.5- to 125-μm multimode fiber is typically 0.7 dB with variances (3 SDs) of ±0.15 dB [7]. With 100- to 140-μm fiber average loss reduces to 0.4 dB.

Figure 5.15 illustrates the process for field assembly of the Thomas & Betts SMA connector. Whereas many other designs require the fiber to be polished (a labor-intensive process) this connector is designed so that the fiber can be simply precision cleaved.

5.2.3 Biconic Connector

A very popular design for multimode and single-mode fibers is a precision moulded and machined plastic connector with a conically tapered ferrule and sleeve alignment mechanism (Fig. 5.16).

The biconic connector uses a cone-shaped plug that has a small hole, precision machined in the center, that accepts a single-mode or multimode fiber. The plugs and hole sizes are sorted to ensure maximum concentricity of the fiber within the connector. Connectors can be polished so fibers butt against each other or short-polished to prevent fiber contact. The biconic alignment technique was developed at Bell Laboratories. The information presented here is from Dorran and EOTec [11, 12].

Connectors are available for 125-, 140-, 250-, and 400-μm fiber outer diameters. Insertion loss with single-mode and multimode fiber is typically 0.6 dB for contacting fibers and 0.7 dB for noncontacting, with variations generally from 0.5 to 1.1 dB. Reflection for fiber-contacting connectors is typically less than 32 dB below transmitted power levels.

5.2.4 ST Connector

The ST connector is a trademark of AT&T. The Thomas & Betts design, shown in Fig. 5.17(a) [7], has a housing that is mated by pushing and rotating the plug cover onto a bayonette-style sleeve housing. Recently an ST-compatible push-pull design has been developed by 3M that is illustrated in Fig. 5.17(b) [13].

A tight tolerance for the fiber hole and ferrule concentricity is achieved by using a precision manufactured ferrule, which fits into a precision mating sleeve. To get a precision concentric hole in the center of a rigid ferrule, several designs and materials are used that permit the precision manufacturing including zirconia ceramic, alumina ceramic, glass capillary in ceramic, glass capillary in stainless steel, glass capillary in plastic, and ARCAP a stainless steel and copper-based material.

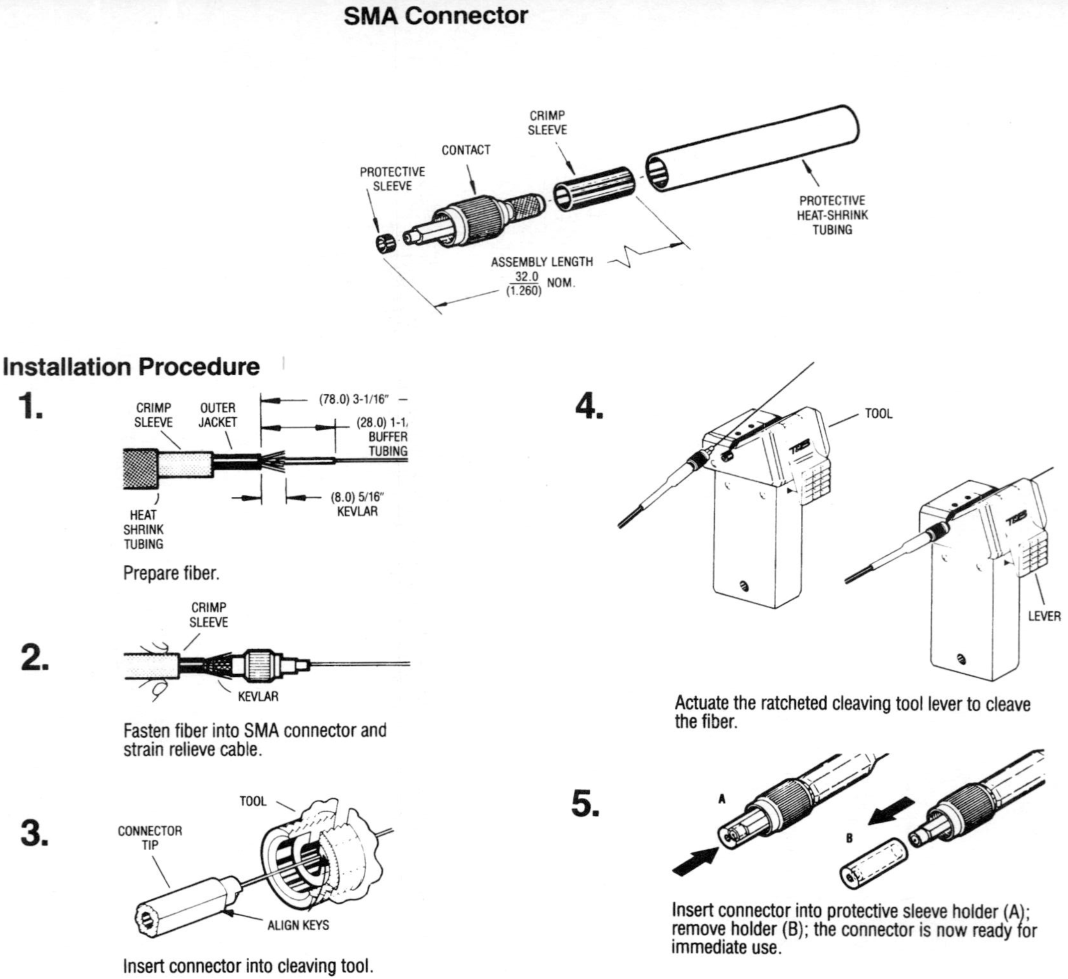

SMA Connector

PROTECTIVE HEAT-SHRINK TUBING

CRIMP SLEEVE

CONTACT

PROTECTIVE SLEEVE

ASSEMBLY LENGTH 32.0 NOM. (1.260)

Installation Procedure

1.

(78.0) 3-1/16"
(28.0) 1-1. BUFFER TUBING
(8.0) 5/16" KEVLAR

CRIMP SLEEVE
OUTER JACKET
HEAT SHRINK TUBING

Prepare fiber.

2.

KEVLAR

CRIMP SLEEVE

Fasten fiber into SMA connector and strain relieve cable.

3.

TOOL

CONNECTOR TIP

ALIGN KEYS

Insert connector into cleaving tool.

4.

LEVER

TOOL

Actuate the ratcheted cleaving tool lever to cleave the fiber.

5.

A

B

Insert connector into protective sleeve holder (A); remove holder (B); the connector is now ready for immediate use.

Figure 5.15. Thomas & Betts installation procedure for SMA connector. (From Ref. 7; copyright

134

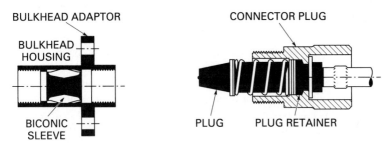

Figure 5.16   Biconic connector design.

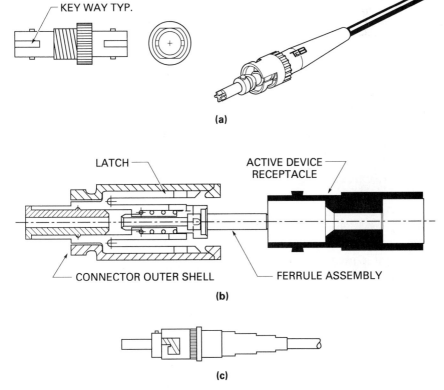

Figure 5.17   ST connectors. (a) Bayonette style (From Ref. 7; copyright Thomas & Betts); (b) self-alignment sequence of push-pull ST compatible connector in relation to bore of active device receptacle (From Ref. 13; copyright 3M); and (c) 6100 hot-melt connector compatible with ST connectors (PC finish). (From Ref. 28; copyright 3M).

In the case of 3M [13] a zirconia ceramic ferrule is used with a proprietary finish that optimizes optical contact and reduces reflection. Thermoplastic components are used within the connector to prevent metal flaking. Ferrules come in sizes with 1-μm variations to accommodate fiber outer-diameter tolerances. 3M also offers a hot-melt connector for quick-mount field connection to multimode fiber [28]. It uses a zirconia ferrule with a hot-melt adhesive inside that holds the fiber once it is cooled. The end is then polished.

Single-mode attenuation is typically about 0.15 to 0.25 dB for ST connectors with maximum attenuation between 0.4 to 0.7 dB depending on manufacturer and end finish [9, 10, 13]. Multimode attenuation averages between 0.05 to 0.15 dB with maximums between 0.1 and 0.3 dB. The hot-melt design achieves a typical attenuation of 0.3 dB [28]. Single-mode reflection is about 40 to 50 dB below the transmitted power level (1/10,000 to 1/100,000) and multimode around 25 dB to 36 dB below.

The ST connector appears to be one of the most popular connector designs, with an installed base on the order of more than 70% of the connectors sold in the United States in 1990 [14]. It is not as compact a design as the SC connector (see later) and only recently has been made into the push-pull design [13] that the SC connector is popular for. It can be factory or field mounted to the fiber. The ST and its clones are field mounted on the fiber using three or fewer pieces, and then polished.

The connector has been or is in process of being accepted by various standards committees including IEC-BFOC, MIL-C-83522, and EIA/TIA-BFOC.

5.2.5 SC Connector

The SC connector is an NTT design aimed at high-density packaging and coupling. Shown in Fig. 5.18, it also provides a tight tolerance for the fiber hole and ferrule concentricity by using a precision-manufactured ferrule.

The NTT specification [13] calls for the following:

- Keyed rectangular design
- Isolated ferrule design that is stable for high-density packaging
- Push-pull coupling
- Single and four-pack designs for high-density

Single-mode attenuation is typically about 0.25 dB for this connector design with variations from 0 to 0.6 dB between matings. Single-mode reflection is about 40 to 50 dB below the transmitted power level (1/10,000 to 1/100,000).

The SC connector is relatively new and therefore represents only about 1% of the installed base in the U.S. but is gaining in popularity quickly because of its push-pull actuation (non-rotating), compact design, and self-aligning features. The ST con-

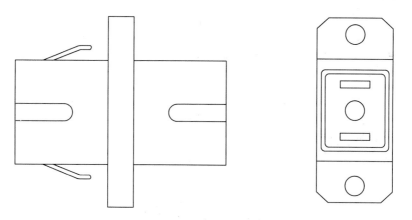

Figure 5.18   SC connector.

nector requires that the ferrule be aligned with the sleeve before inserting, whereas the SC connector does not. Both the ST and SC connectors are expected to account for about two-thirds of the market by 1995 [17].

It can be factory or field mounted to the fiber. The original SC design called for eight pieces and then polishing to field mount the connector on the fiber.

The connector has been proposed for acceptance by various standards committees including IEC and EIA/TIA, and may be adopted as standard by some Bell Operating Companies.

5.2.6 FC Design

The FC design is a ferrule-in-sleeve concept but is designed so that the plug housing screws onto the sleeve housing. As shown in Fig. 5.19, the plug is keyed and the sleeve housing notched so that the ferrule does not rotate.

As with the other precision ferrule-in-sleeve designs, various ferrule designs are used to achieve the precision concentric fiber acceptance hole. In the AMP design the sleeve has a spring-loaded clip that aligns the outer surface of the ferrules and holds them tightly in alignment.

The ferrule and fiber can come with an end finish that permits physical contact of the fibers, thus reducing insertion loss and reflection dramatically. This is known as a physical contact (PC) connector. Because the ferrule is keyed it cannot rotate such that the contacting fibers scratch each other.

For single-mode fiber the insertion loss is typically 0.4 to 0.7 dB with maximum loss variations (3 SDs) between 0.8 to 1.0 dB depending on manufacturer [12–15]. With the PC type, mean loss can be reduced to 0.15 dB. Multimode insertion loss is typically from 0.12 to 0.6 dB with maximum variations of 1.0 dB, depending on fiber core size and manufacturer [7, 20, 21].

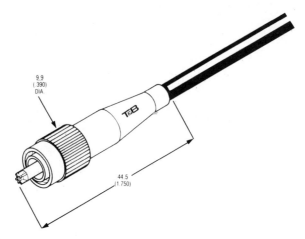

Figure 5.19   FC connector design. (From Ref. 7; copyright Thomas & Betts.)

5.2.7 FDDI Connector

The FDDI connector, illustrated in Fig. 5.20, is specified as part of the ANSI FDDI Physical Layer Medium Dependent standard X3T9/86-71 [22]. It is called a Media Interface Connector (MIC). The primary function of the connector is to connect the optical transmission fiber with another transmission fiber or with an optical port on an FDDI component such as a receiver, a transmitter, or a bypass switch. The plug has latch points that mate with latches on the body of the receptacle, and are polarized mechanically. The connector contains two ferrules, one for each fiber of a duplex cable. The ferrules are mounted in such a way as to give them resilient movement during the mating process. The standard permits any design so long as it is compatible with the geometrical requirements given in the standard.

The connectors also have four key configurations for the four types: MIC A (primary in/secondary out), MIC B (secondary in/primary out), MIC M (used with a concentrator), and MIC S (used with the single attachment station).

Ferrule size, shape, and mating force are defined in the standard.

FDDI connectors are anticipated to only be as popular as the number of FDDI systems installed, and will be in competition with ST and SC duplex designs.

5.2.8 Ferrule End Finish

Where connectors have polished ferrules, various end finishes can be applied. The three most common are illustrated in Fig. 5.21, the GAP, flat, and PC finish. Although a GAP finish protects the fiber surfaces from scratching each other, the air gap creates a loss component of 0.35 dB owing to reflection from the glass to air to glass refractive index interfaces. This reflection also can affect the performance of some systems by scattering light in the reverse direction down the fiber.

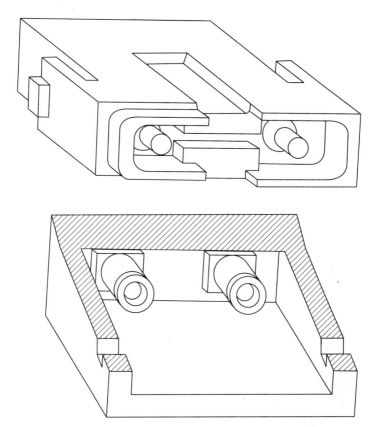

Figure 5.20   FDDI connector configuration.

The lower reflection of the flat and PC finishes are also illustrated in Fig. 5.21.

5.3 COUPLERS

In the systems that have been discussed so far only two terminals are used, a transmitting terminal and a receive terminal connected to opposite ends of a fiber span. It is possible, by using optical couplers, to attach more than one set of transmit and receive terminals to a single fiber rather than running a separate fiber or cable for each transmit-receive pair.

5.3.1 Coupler Characteristics

The most common application of this technology is with LANs, whereby a common fiber carries the multiplexed signals from multiple terminals placed at various locations served by the LAN. Access to the LAN is made through optical couplers that

STRAIGHT FERRULED CONNECTOR END SHAPES

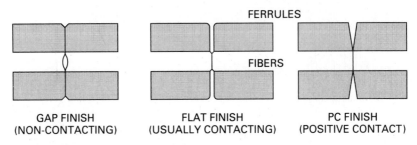

FERRULES

FIBERS

| GAP FINISH | FLAT FINISH | PC FINISH |
| (NON-CONTACTING) | (USUALLY CONTACTING) | (POSITIVE CONTACT) |

REFLECTIVITY
FLAT/FLAT vs FLAT/PC vs PC/PC

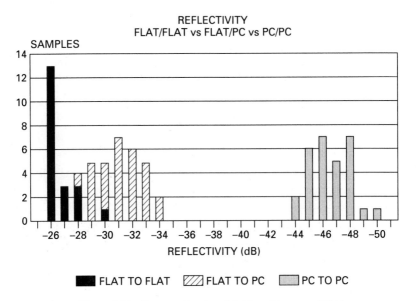

Figure 5.21   Connector ferrule end finishes. (Courtesy of 3M.)

divert part of the signal power on the LAN fiber to each receiver and couple power from each terminal transmitter onto the fiber.

Optical couplers are also used to combine power from many sources of different wavelengths onto a single fiber to permit transmission in both directions simultaneously on a single fiber, and other such applications where a portion of the power is to be directed to or from the core of the fiber. Some of the more common coupler configurations are illustrated in Fig. 5.22 and discussed later.

Couplers affect the performance of a fiber-optic system by splitting and diverting power. The amount of power arriving at any receiver is affected by the amount of total power that each coupler in line has diverted to the path of that receiver. This

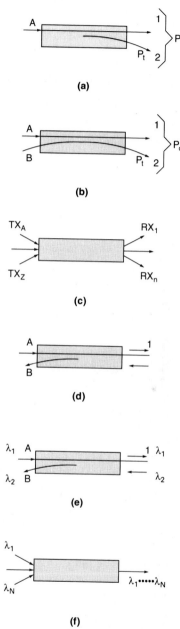

Figure 5.22 Common optical coupler configurations.

power splitting factor is called the coupling ratio (CR). Couplers also reduce total power through internal loss mechanisms known as excess loss (Le), and through insertion loss of any associated connectors (Lc). The total insertion loss of a coupler is a combination of the three:

$$\text{Coupler insertion loss (dB)} = \text{Le} - 10 \log \text{CR} + 2 \text{ Lc}$$

where

$$\text{CR} = \frac{\text{power coupled to receive port}}{\text{total power out to all ports}} = \frac{P_t}{P_{out}}$$

Performance of typical couplers is given in Table 5.2 [1].

5.3.2 Tree and Branch Coupler

The tree and branch coupler, also known as a T coupler, is illustrated in Fig. 5.23(a) and (b). The three-port coupler design permits power from a transmit fiber (port A) to be split to two ports (1 and 2) in some coupling ratio, usually 1:1 or 1:n where n is some fraction. In a four-port coupler, power from a second transmit fiber (port B) can be coupled onto the main fiber bus (port A to 1), while an equal amount of power is coupled off (at port 2). Some applications of the tree and branch couplers are illustrated in Fig. 5.23. Note that although these couplers are sometimes used as a power combiner or for duplex transmission, a 3-dB (50%) splitting loss is suffered as a result. This splitting loss can be eliminated by using two wavelengths and a wavelength coupler, or by a directional coupler.

5.3.3 Star Coupler

The star coupler (see Fig. 5.22[c]) is a multiport port coupler that permits power from one of N transmit ports (A through Z) to be split equally to each of N receive ports (1 through N). In this case the coupling ratio is 1/N. One of the most common applications of the star coupler is illustrated in Fig. 5.24.

5.3.4 Directional Coupler

The directional coupler (see Fig. 5.22[d]) is a three-port coupler. It permits power to be transmitted in one direction down a fiber (port A to port 1), whereas power at the same wavelength is received from the other direction and routed from port 1 to port B. Unlike the T coupler, no power splitting loss is experienced, only the small excess loss and connector insertion losses.

For single-mode fiber, directional couplers are generally fabricated using fused fibers, although some single-mode integrated-optics flat-substrate couplers achieve

TABLE 5.2 TYPICAL PERFORMANCE OF FIBER COUPLERS

Design class	No. of ports	Coupling ratio (CR)	CR tolerance (±%)	Excess loss (dB)	Uniformity (dB)	Isolation directivity (−dB)	Polarization sensitivity (%)
2 × 2	2	0.5	2.0–15%	0.07–1.0	0.1–0.2	−40 to −55	0.1–6
Single mode	2	0.25	or	0.07–1.0	0.1–0.2	−40 to −55	0.1–6
WIC or WDC	2	0.1	0.03 dB/nm	0.07–1.0	0.1–0.2	−40 to −55	0.1–6
2 × 2	2	0.5	5–10%	<1	0.5	−35 to −40	
Multimode	2	0.25	5–10%	1–2	0.5	−35 to −40	
	2	0.1	5–10%	1–2	0.5	−35 to −40	
	2	0.0625	5–10%	1–2	0.5	−35 to −40	
$N \times N$ WIC	3	0.33	10%	<0.5	0.1–0.2		1
Star coupler	4	0.25	10%	0.5–2.0	0.5–0.6		
	8	0.125	10%	<2.5	0.6		
	16	0.0625	10%	<6.0	0.6		
	32	0.03125	10%	<8.0	0.6		
Directional 2 × 2 WIC	2	1		0.5–1.0	5–10	−40 to −50	<5

Design class	No. of ports	Wavelength spacing (nm)	Wavelength channels (μm)	Excess loss (dB)	Wavelength sensitivity (± %)	Isolation directivity (−dB)	Polarization sensitivity (%)
Wavelength Multiplexing							
2 × 2 WDM	2	50–200	1.3,1.5	0.4–0.8	0.1 to 0.4%/nm	−16 to −28	<2.5%
Wavelength dependent						(−59 to −60 w/filter)	
$N \times N$ WDM	3	200–200	1.2, 1.3, 1.5	4.0		−25	
Single mode	6	20–30	1.3 or 1.5	2–3		−25	
	8	20–30	1.3 or 1.5	2–3		−25	
$N \times N$ WDM	3	100–200	1.2, 1.3, 1.5	3.5		−25	
Multimode			or 1.1, 1.2, 1.3				

WIC = Wavelength independent coupler
WDC = Wavelength dependent coupler

the same function. For multimode fiber, directional couplers are generally fabricated with some bulk components using beam dividers and reflective surfaces to transmit light through in one direction while reflecting receive light coming from the other direction onto the receive fiber or detector.

Some applications of directional couplers are illustrated in Fig. 5.25.

5.3.5 Wavelength-Dependent Coupler

Wavelength-dependent couplers are illustrated in Fig. 5.22(e). The three-port coupler design permits: (a) power to be transmitted from two sources at different wavelengths onto the same fiber (ports *A* and *B* onto port 1); or (b) power to be transmitted in one

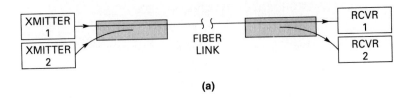

(a)

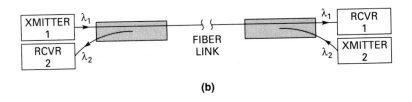

(b)

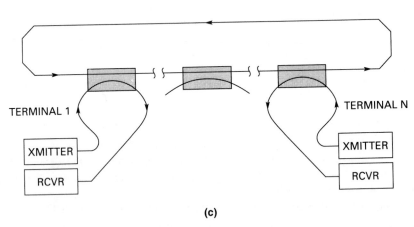

(c)

Figure 5.23 Applications of the tree and branch coupler. (a) Power combiner and splitter, (b) duplex multimode transmission link, and (c) LAN "tap."

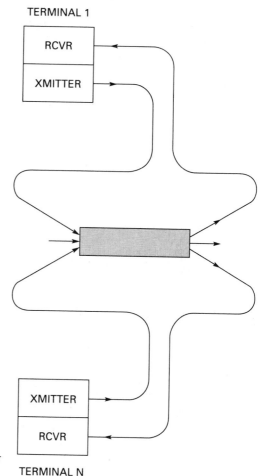

TERMINAL 1

Figure 5.24 LAN application of the star coupler.

TERMINAL N

direction down a fiber (port *A* to port 1), whereas power at a separate wavelength is received from the other direction and routed from port 1 to port *B*. Unlike the T coupler that splits power, this coupler diverts power of a particular wavelength in a particular direction, much like a prism does. No power splitting loss occurs; therefore, only a small internal excess loss and any connector insertion loss is experienced.

For single-mode fiber these couplers are generally fabricated using fused biconically tapered fibers. For multimode fiber, the couplers are generally fabricated with some bulk components using beam dividers and wavelength-sensitive reflective surfaces to transmit light through in one direction while reflecting receive light coming from the other direction onto the receive fiber or detector.

Some applications of the wavelength-dependent couplers are illustrated in Fig. 5.26.

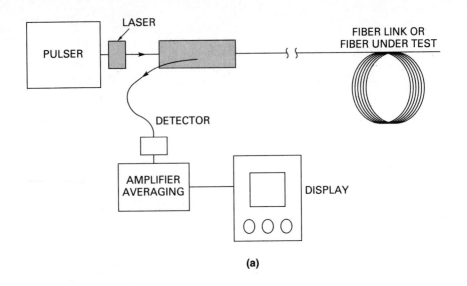

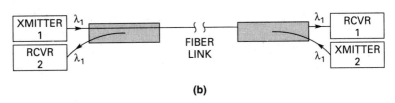

Figure 5.25 Applications of directional couplers. (a) Optical-time-domain reflectometer and (b) duplex single-mode transmission link (same wavelength in both directions).

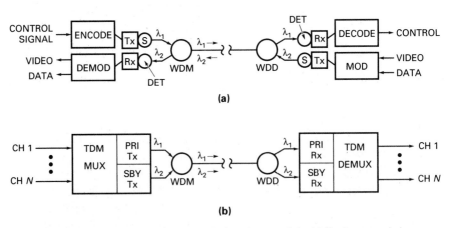

Figure 5.26 Applications of wavelength-dependent couplers. (a) Duplex transmission link, and (b) wavelength multiplexer for doubling fiber transmission capacity or accommodating protection channel electronics. WDM = wavelength-division multiplier coupler; WDD = wavelength-division de-multiplexer coupler; PRI = primary; SBY = stand by; RX = receiver; TX = transmitter.

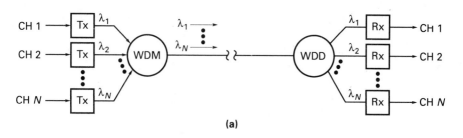

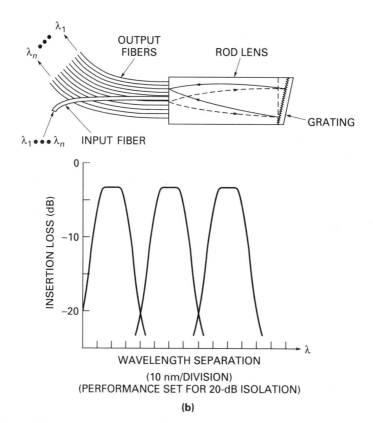

Figure 5.27 Wavelength-division multiplexing. (a) Applications of wavelength-division multiplexing, and (b) wavelength-division multiplexing grating coupler.

5.3.6 Wavelength Multiplexing

Wavelength multiplexing and demultiplexing is illustrated in Fig. 5.22(f). The multiport coupler design permits power to be transmitted from multiple sources at different wavelengths onto the same fiber (ports A through Z onto port 1). It also permits multiple communications channels composed of power of multiple wavelengths to be separated by wavelength, each channel being diverted to a separate receive fiber (port A to ports 1 through N). This coupler does not split power, but, like a prism, diverts power of a particular wavelength in a particular direction. No power-splitting loss occurs; only an internal excess loss and any connector insertion loss is experienced.

For single-mode or multimode fiber these couplers are generally fabricated using bulk devices. Diffraction gratings may be used within the demultiplexer, for example, to diffract the different wavelengths of light entering the coupler to positions where separate receive fibers are attached.

The application of wavelength-division multiplexing is illustrated in Fig. 5.27 on page 147.

REFERENCES

1. R. Hoss, *Fiber Optic Communications Design Handbook* (Englewood Cliffs, N.J.: Prentice Hall, Inc., 1990), pp. 122–127.

2. "Introduction to Fiber Optics and AMP Fiber-Optic Products," AMP HB 5444, AMP Inc., n.d.

3. Siecor product literature, "M90 Microprocessor Controlled Fusion Splicer," 1992.

4. Based on information obtained in PSI Telecommunications Limited product literature on the Lightlinker fiber optic splice system.

5. GTE elastomeric splice, manufacturers literature and test data.

6. AT&T Practice 640–252–176, Single Mode Lightguide Cable Rotary Splicing, AT&T Technologies, May 1985.

7. Thomas & Betts Electronics Division, product literature FOC-1, "Fiber Optic Interconnect Systems."

8. "Introduction to Fiber Optics and AMP Fiber-Optic Products," AMP HB 5444, AMP Incorporated, n.d.

9. AMP data sheet 78-513, "AMP Optimate Small Fiber Connector."

10. Terry Brown, "Low Cost Connectors for Single Optical Fibers," presented at Electronic Components Conference, Anaheim, California, April 24–26, 1978.

11. Dorran Photonics, Inc., data sheet 1–10, "Multimode Biconic Jumper Assembly."

12. EOTec Corp. data sheet, "Fiber Optic Cable Assemblies, Biconic Type."

13. 3M Product Bulletins "Push-Pull Connector—ST" and "SC Connector."

14. "U.S. Markets for Fiberoptic in Point-to-Point Data Communications," Kessler Marketing Intelligence.

15. Composite data from various manufacturers.

16. 3M proposal to the TIA TR-41.8.1 subcommittee on building wiring, September 10, 1991.

17. "SC/ST Dominance," *Lightwave*, Sept. 1991.

18. Amphenol data sheet, "Single mode FC cable assemblies," 942 series.

19. Sumitomo Electric data sheet, "OPTOPIA Optical Fiber Connector."

20. OFTI newsletter, *Top Score for OFTI's New All Metal STC Connector*, Aug. 1987.

21. EOTec data sheet, "Fiber Optic Cable Assemblies—FC NTT Type."

22. ANSI Standard, "FDDI Physical Layer Medium Dependent (PMD)," X3T9/86-71, X3T9.5/84-48 Rev. 8, 7-1-88.

23. W. S. Hudspeth, "Fiber Optic Connectors—Still a Budding Technology," *Electro-Optical Systems Design,* Oct. 1978, p. 48.

24. AT&T product literature, "Universal Fiber Optic Closure, F-83AK8530,8531."

25. 3M product literature, "Fibrlok Optical Fiber Splice."

26. 3M technical report, "Fibrlok Optical Fiber Splice," Oct. 1989.

27. 3M product literature, "6100 Hot-Melt Fiber Optic Connector."

28. Northern Telecom product literature, "Optical Fiber Splice Organizer."

6

Detectors and Receivers

At the receiver end of a fiber optic system, lightwaves are first converted to an electrical current and then amplified and converted to a voltage signal by the first amplifier stage known as a preamplifier. The resultant signal is then conditioned for further application in a different fashion depending on whether it is analog or digital. If analog, the preamplifier is generally followed by more linear amplifier stages and perhaps some automatic gain control. If digital, the signal is amplified and then fed to a threshold detector and timing electronics that decode it into binary signals.

6.1 PHOTODETECTORS FOR FIBER OPTICS

Semiconductor light sensors (photodetectors) are used to convert the optical energy to electrical current. The detectors most commonly used in fiber optics are positive-intrinsic-negative (PIN) photodiodes and avalanche photodiodes (APDs).

Although phototransistors and photodarlington transistors have a higher gain than PIN photodiodes and have been used for some fiber applications, they have low bandwidth (150 kHz) and relatively high noise. We will therefore concentrate on PIN and APD devices in this chapter.

The difference between an APD and a PIN detector is that an APD has internal gain. In an APD multiple electrons are created when a photon is absorbed, whereas in a PIN detector a maximum of one electron is created per photon. As a result of this added gain APDs have a greater sensitivity than PIN detectors. In the 850-nm range, where silicon APDs have gains on the order of 100, this sensitivity difference is 10 to 20 dB greater than PIN detectors. At longer wavelengths, practical device limitations hold the gain to between 10 and 30, so the advantage is not so great.

In general, APDs cost 5 to 10 times more than PINs. APDs also require bias voltages that are higher than PINs. In the 850-nm range, silicon APDs require bias voltages of around 250 V to obtain the optimum gains, whereas PIN devices only require 10 to 50 V. This implies that a high voltage (very low current) supply must be provided with an APD receiver, which further increases the cost differential. At longer wavelengths (1300 to 1550 nm) the bias requirement for APDs is 20 to 30 V, whereas that of PIN devices is 5 to 15 V.

6.2 SIGNAL-TO-NOISE RATIO IN OPTICAL RECEIVERS

For use in fiber optics, photodetectors must be highly sensitive (responsive) to weak light signals yet have sufficient bandwidth or speed of response to handle the incoming signals. At the same time the detectors must be relatively immune to changes in temperature and must add a minimum of noise to the circuit.

The response of a photodetector is expressed by the term *responsivity* (*R*), which is the ratio of output current to the incident optical flux. It is measured in amperes per watt.

One of the main factors that determine sensitivity is the noise in the circuit. In any fiber-optic system there is a certain amount of spurious, unwanted energy called *noise*. It can be caused by electromagnetic interference, crosstalk, and other phenomena. In light-related devices, there is always a small trickle of current, called the *dark current,* which flows when there is no light in the circuit. Thus, any circuit has a prevailing noise level, a certain level of unwanted electrical energy. Additional bursts of energy will cause noise above the prevailing level [1] as shown in Fig. 6.1.

If the signal's strength is less than the prevailing noise level, the signal will be lost in the noise. At the same time, extra bursts of noise may be interpreted as part of the signal, as a pulse, for example [1].

Two terms are used to describe the relationship between noise and the signal: *signal-to-noise ratio* (SNR or S/N) and *bit-error rate* (BER). SNR expresses how much stronger the signal is than the noise at any point in the system. BER is the equivalent measurement used at the output of a digital receiver for the number of bits decoded in error versus the total number of bits received. In an optical system, SNR is generally referenced to signal current and noise current at the photodetector. It is, however, more practically defined in terms of the ratio of signal power to noise power at the optical input of the photodetector or electrical output of the receiver preamplifier, or signal voltage to noise voltage at the preamplifier output. It is generally expressed as

$$SNR(dB) = 10 \log(\text{signal power} / \text{noise power})$$

or

$$SNR(dB) = 20 \log(\text{signal voltage} / \text{noise voltage})$$

SNR is also defined in different ways depending on the waveform. In analog

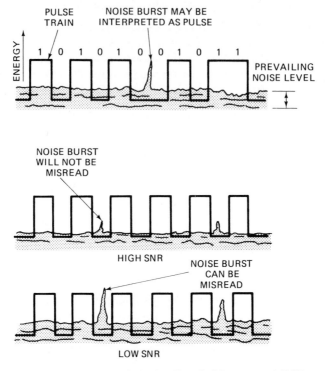

Figure 6.1 Electrical noise. (From Ref. 1; courtesy of AMP.)

transmission systems it is more convenient to define it in terms of root-mean-square (rms) signal to rms noise power. An rms measurement is similar to an average level, which is the most convenient measurement to use when the signal and noise level is constantly varying. In digital transmission, where the peak level (the "on" state) of the signal is a known, it is most convenient to define SNR as peak signal voltage to rms noise voltage level.

SNR is statistically related to BER. Referring to Fig. 6.2 [2], the signal is reconstructed at the receiver by presenting the composite received signal, with the noise distortion on it, to the threshold detector of a digital receiver. If noise causes the composite signal to cross the decision threshold when it is not intended to, the pulse may be interpreted by the threshold detector in error. The probability that this will happen under certain noise conditions is described by BER. The lower the SNR, the higher the probability that the signal pulses will be interpreted in error.

The relationship between SNR and BER is different but predictable for various encoding and detection schemes and is well known in receiver design [2]. For example, for a digital fiber-optics receiver, using a simple decision threshold and NRZ encoding, to produce a BER of 10^{-9} (1 decoding error in every 1 million bits transmitted), theory dictates that a SNR (peak signal voltage to rms noise voltage) of 21.6 dB

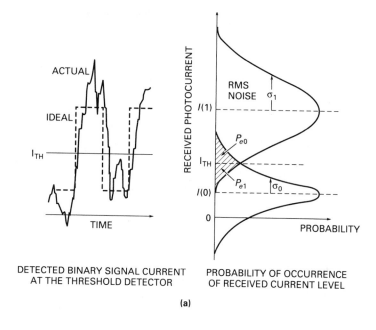

ACTUAL

IDEAL

I_{TH}

TIME

DETECTED BINARY SIGNAL CURRENT
AT THE THRESHOLD DETECTOR

RECEIVED PHOTOCURRENT

RMS
NOISE σ_1

$I(1)$

I_{TH}

P_{e0}

P_{e1} σ_0

$I(0)$

0

PROBABILITY

PROBABILITY OF OCCURRENCE
OF RECEIVED CURRENT LEVEL

(a)

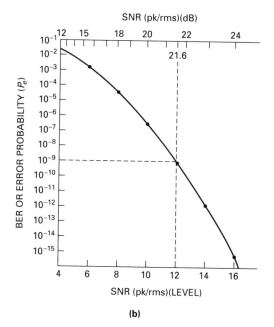

SNR (pk/rms)(dB)

21.6

BER OR ERROR PROBABILITY (P_e)

SNR (pk/rms)(LEVEL)

(b)

Figure 6.2 The relationship of signal to noise ratio, at the receiver threshold detector, to bit-error rate. (a) Error probability for binary states. $I(0)$ = mean photocurrent for a binary 0, $I(1)$ = mean photocurrent for a binary 1, I_{TH} = decision threshold level, σ_0 = rms signal variance owing to noise for a binary 0, σ_1 = rms signal variance due to noise for a binary 1, P_{e0} = probability that a 1 was detected when a 0 was sent, and P_{e1} = probability that a 0 was detected when a 1 was sent. (b) Relationship of BER to received SNR (pk/rms) (From Ref. 2; copyright 1990 Prentice Hall, Inc.)

or greater is required. This relates to a peak signal voltage 12 times greater than the rms noise voltage level.

Note that noise is a direct function of the bandwidth in which it is measured, and thus related to the bandwidth requirements of the signal. The greater the bandwidth (MHz) or bit rate (Mb/s) of the signal being transmitted, the greater the bandwidth required of the receiver, and thus the more noise produced by the photodetector and receiver. High-speed systems thus require more power, and therefore lower attenuation components, than low-speed systems to maintain the same SNR or BER.

S/N measurements in an optical system are generally referenced to the receiver, either at the input or the output of the preamplifier stage, before any further analog or digital signal processing. Figure 6.3 illustrates the sources of signal and noise components that make up SNR.

The signal is the optical power swing created by the original information signal current that drives the LED or ILD. When it reaches the receiver it is reduced in amplitude by the losses in the fiber, splices, connectors, and couplers, and it is somewhat distorted by the bandwidth limitations of the fiber and optical source. The noise present on the optical signal at this point is generally negligible, but can in some cases be important. This noise consists of unwanted signal components (such as harmonic distortion), intersymbol interference in digital signals, spurious interference, and perhaps what is known as "modal noise" created in fibers and optical sources [2].

In most fiber systems, most noise is created at the receiver, from a combination of photodetector noise and noise generated within the preamplifier transistors and resistors. The photodetector noise is created from a combination of (1) unwanted current, called "dark current" that flows through the photodetector as a result of having a bias voltage applied (similar to thermal noise in a resistor); and (2) noise that results in the process of converting optical power to electrical current in the photodetector (called "quantum" noise). In PIN detectors dark current noise predominates, whereas in APDs quantum noise is the larger.

Amplifier noise is created from two sources as well: (1) thermal noise, which results from random fluctuations in current flowing through the resistive elements of the receiver; and (2) transistor noise caused by various noise-generation mechanisms in the gain process of the transistors. Low-noise amplifier design is a matter of increasing input resistance to limit thermal noise and using very low noise transistors.

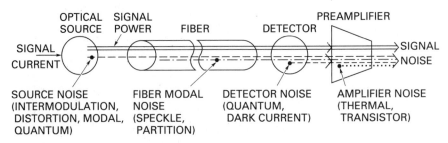

Figure 6.3   Noise sources in a fiber-optic system.

The noise generated by the detector and receiver defines the sensitivity of the receiver. This receiver sensitivity generally dictates the performance limitations for the entire fiber system. Receiver sensitivity is related to S/N, within the bandwidth of the signal being transmitted. It is defined as that minimum optical signal power level at the detector required to produce the SNR or BER at the output of the receiver, as specified for proper system operation. Low-noise receiver design is a matter of proper component selection and matching of the photodetector with preamplifier design for lowest-noise performance [2, 3].

6.3 PIN PHOTODIODES

In the photodiode family, there are two basic types: depletion layer and avalanche. The two types are similar except that the avalanche type has a built-in gain mechanism.

Of the various depletion-layer types, only the PIN diode (Fig. 6.4) is significant for use in fiber optics. In fact, it is the most common light sensor in such circuits.

Figure   6.4   RCA   C-30920E   PIN   photodiode. (Courtesy of RCA.)

As shown in Figs. 6.5 or 6.6, the PIN photodiode has a layer of undoped or intrinsic (*I*) material sandwiched between a layer of positively (*P*) doped and negatively (*N*) doped material. Light enters the diode through a tiny "window" which is about the size of the fiber core.

In effect, the photodiode works just the opposite from an LED. Photons that fall on the carrier-void depletion region (Fig. 6.7) create carriers. Light-generated carriers

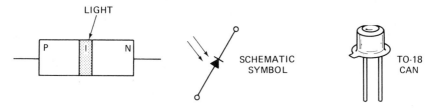

Figure 6.5   PIN photodiode. (From Ref. 1; courtesy of AMP.)

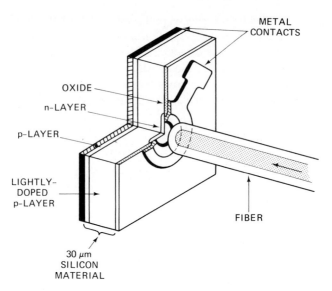

Figure 6.6   Cut away view of PIN photodiode. (From Ref. 4; copyright 1975 Bell Laboratories RECORD.)

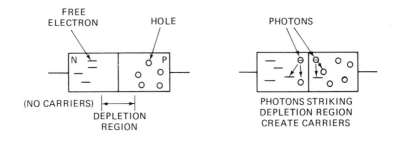

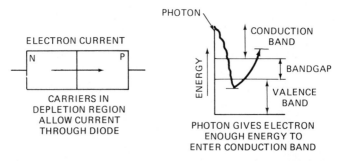

Figure 6.7   Operation of a photodiode. (From Ref. 1; courtesy of AMP.)

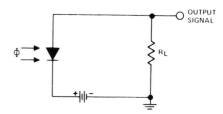

Figure 6.8 Basic solid-state photodiode circuit. (From Ref. 4; courtesy of RCA.)

Figure 6.9 MFOD104F PIN photodiode. (From Ref. 5; copyright 1979 Motorola, Inc.)

in the depletion region allow current to flow through the diode. Notice that the photons are absorbed by the electrons in the valence band. This allows electrons to break free of the valence band and enter the conduction band. The result is a free electron and a hole, both carriers of current. When the light is removed, the depletion region is restored and current stops [1].

The PIN photodiode provides no gain or amplification. For every photon captured in the intrinsic layer, an electron-hole pair is set flowing as current. Because amplifiers can be added after the diode, the lack of amplification is not always a problem [1]. As shown in Fig. 6.8, the PIN requires a bias voltage, which may be as low as 5 V.

Tables 6.1 and 6.2 give the characteristics of a silicon PIN diode as might be used in a relatively low-cost, low-speed data link application. Note that responsivity of the detector itself, which would normally be around 0.5 to 0.6 A/W, is obviously derated for reflectance and coupling loss to the surface owing to the packaging design. The package design, which includes an integral connector, is illustrated in Fig.

TABLE 6.1 MAXIMUM RATINGS FOR MOTOROLA MFOD104 PIN PHOTODIODE[a]

Rating	Symbol	Value	Unit
Reverse voltage	V_R	100	V
Total device dissipation at $T_A = 25°C$ Derate above 25°C	P_D	100 0.57	mW mW/°C
Operating temperature range	T_A	−30 to +85	°C
Storage temperature range	T_{stg}	−30 to +100	°C

Source: Ref. 6 copyright 1979 Motorola Inc.
[a]$T_A = 25°C$ unless otherwise noted.

TABLE 6.2 ELECTRICAL AND OPTICAL CHARACTERISTICS FOR MOTOROLA
MFOD104F PIN PHOTODIODE[a]

Characteristic	Symbol	Min.	Typ.	Max.	Unit
Electrical characteristics					
Dark current	I_D	—	—	2.0	nA
$(V_R = 20 \text{ V}, R_L = 1.0 \text{ M}, H = 0)$					
Reverse breakdown voltage	BV_R	100	200	—	V
$(I_R = 10 \text{ μA})$					
Forward voltage	V_F	—	0.82	1.2	V
$(I_F = 50 \text{ mA})$					
Total capacitance	C_T	—	—	4.0	pF
$(V_R = 5.0 \text{ V}, f = 1.0 \text{ MHz})$					
Noise equivalent power	NEP	—	50	—	$\text{fW}/\sqrt{\text{Hz}}$
Optical characteristics					
Responsivity at 900 nm	R	0.15	0.40	—	μA/μW
$(V_R = 5.0 \text{ V}, P = 10 \text{ μW}^{b})$					
Response time at 900 nm	t_{on}, t_{off}				ns
$V_R = \quad 5.0 \text{ V}$		—	6.0	—	
12 V		—	4.0	—	
20 V		—	2.0	—	
Numerical aperture of input port, 3.0 dB	NA	—	0.48	—	—
[200-μm (8-mil)-diameter core]					

Source: Ref 6; copyright 1979 Motorola Inc.
[a]$T_A = 25°C$
[b]Power launched into optical input port. The designer must account for interface coupling losses.

6.9. For high-speed devices an optical fiber pigtail is generally integrally coupled to the chip to reduce reflectance and thus achieve higher responsivities. Also the chip can be made smaller (the size of the fiber core), and thus it will have lower capacitance and faster rise time.

Table 6.3 illustrates the performance of typical optical detector product used for higher-speed applications [7]. The performance of silicon, germanium, and InGaAs is compared.

Some of the terms in Table 6.3 that were not previously defined are defined as follows:

1. *Responsivity (R).* The output of the diode in amps divided by the incident optical power in watts:
   a. *Chip.* The responsivity assuming all incident light is coupled into the chip. Fig. 6.10 illustrates the responsivity vs. wavelength for PIN detectors of various materials.
   b. *Coupled.* The basic responsivity of the chip derated to account for coupling loss, a portion of which is due to surface reflection from the chip.
2. *Quantum efficiency.* Average number of electrons emitted divided by the average number of incident photons.

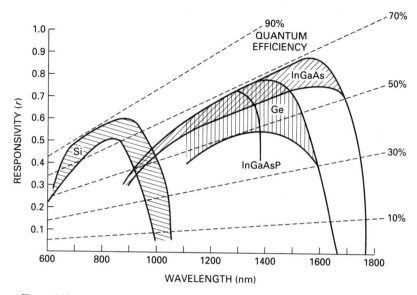

Figure 6.10   PIN photodetector responsivity characteristics. (From Ref. 7; copyright 1990 Prentice Hall, Inc.)

TABLE 6.3 TYPICAL PERFORMANCE OF DETECTION PRODUCT[a]

Parameter	Unit	Silicon		Germanium		InGaAS	
		PIN	APD	PIN	APD	PIN	APD
Wavelength Range	nm	400–1100		800–1800		900–1700	
Peak	nm	900	830	1550	1300	1300 (1550)	1300 (1550)
Responsivity:							
Chip	R	0.6	77–130	0.65–0.7	3–28	0.63–0.8 (0.75–0.97)	
Coupled	R	0.35–0.55	50–120	0.5–0.65	2.5–25	0.5–0.7 (0.6–0.8)	
Quantum efficiency	%	65–90	77	50–55	55–75	60–70	60–70
Gain	G	1	150–250	1	5–40	1	10–30
Excess noise factor	x	—	0.3–0.5	—	0.95–1	—	0.7
K_{eff}	k	—	0.02–0.08	—	0.7–1	—	0.3–0.5
Bias voltage	–V	45–100	220	6–10	20–35	5	<30
Dark current							
Nonmult	NA	1–10		50–500		1–20	1–5
Mult	NA	—	0.1–1.0	—	10–500	—	1–5
Capacitance	pF	1.2–3	1.3–2	2–5	2–5	0.5–2	0.5
Rise time	ns	0.5–1	0.1–2	0.1–0.5	0.5–0.8	0.06–0.5	0.1–0.5

[a]The author wishes to acknowledge Dr. Tran Muoi for some of the information on which this table was developed.

3. *Capacitance (Cd).* The capacitance of the packaged chip. The higher the capacitance, the slower the response and the noisier the receiver. Receiver noise is a function of capacitance partly because lowering of input resistance is needed to compensate (to maintain receiver bandwidth) as capacitance increases, and lowering of resistance increases thermal noise. Capacitance generally decreases with bias voltage.

4. *Rise time (tr).* The time it takes the photodiode current to rise from 10% to 90% maximum level for an instantaneous increase in input optical power. It is related to the speed at which electrons are created and flow across the chip. It generally increases with bias voltage.

The PIN detector exhibits photocurrent noise *(In)* associated with (1) dark or leakage current *(Id)*, and (2) quantum noise that is due to the random nature of electron formation and is therefore a function of the optical power *(P)* incident on the device.

$$In = 2\,e\,[R\,P + Id]\,BW$$

where *BW* is the noise bandwidth of the receiver (usually assumed equivalent to the signal bandwidth requirement), and *R* is the responsivity.

6.4 AVALANCHE PHOTODIODES

In contrast with the PIN photodiode, the APD (Fig. 6.11) has internal gain. This gain is a result of impact ionization, which occurs at high reverse-bias voltages, near the breakdown voltages. During impact ionization a free electron or hole can gain sufficient energy to ionize a bound electron. The ionized carriers cause further ionizations, leading to an avalanche of carriers [3, p. 167].

As shown in Fig. 6.12, the standard APD is a "reach-through" diode. The depletion region has a wide drift region and a narrow multiplying region, as indicated in

Figure 6.11 RCA C30904E avalanche photodiode. (Courtesy of RCA.)

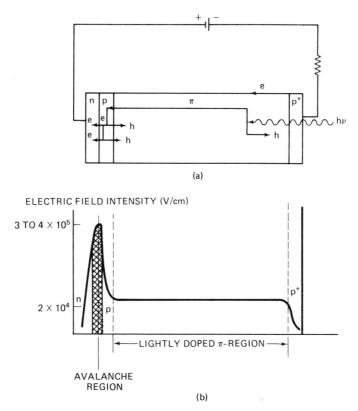

Figure 6.12   Operation of an avalanche photodiode. (Reprinted with permission from Ref. 3, p. 16 [*Electronic Design,* Vol. 28, No. 9]; copyright Hayden Publishing Co., Inc., 1980.)

the electric field profile. Photons are absorbed in the lightly doped π-region, whereas photogenerated carriers cause impact ionization in the avalanche region [3].

When the peak electric field is 5% to 10% below the avalanche breakdown field, the doping levels of the p and n avalanche regions allow the diode depletion layer to "reach through" to the low-doped π-region [3].

The APD has additional noise sources to those of the PIN photodiode, because the internal gain mechanism also amplifies some of the noise sources as well as the signal [7]. The APD exhibits photocurrent noise (In) associated with (1) unmultiplied dark current (Idu) that leaks across the surface of the device and is not subjected to gain; (2) multiplied dark current (Idm) that flows through the device and is multiplied by the gain mechanism (G); (3) quantum noise that is both a function of the optical power (P) incident on the device as well as the gain mechanism (G); and (4) an excess noise factor (F), which is due to randomness in the gain mechanism and therefore multiplies noise slightly more than signal.

$$In = 2 \, e \, [(R \, P + Idm) \, G^2 F + Idu] \, BW$$

where BW is the noise bandwidth of the receiver (usually assumed equivalent to the signal bandwidth requirement), and R is the unmultiplied responsivity.

Figure 6.13 illustrates the typical multiplied responsivities ($R \, G$) of APDs. Figure 6.14 illustrates the responsivity characteristics with gain and temperature variation for a silicon APD device as might be used for a short-wavelength application. Note that gain is strongly influenced by chip temperature. If at room temperature (25°C) the bias voltage is set at 350 V to achieve a multiplied responsivity of 100 A/W, and as the temperature increases to 40°C, the responsivity will decrease to 40 A/W. As a result there will be a 60% reduction in signal level.

To overcome this problem (a problem only APDs have) the receiver must compensate by increasing bias voltage. Generally this is done by monitoring the average level of the signal and if it decreases, bias voltage is increased until the level is recovered, as illustrated in Fig. 6.15 [7]. This is known as automatic gain control (AGC). AGC implies that a constant signal level is present; therefore, the use of APDs has implications in signal encoding as well.

Table 6.3 presents the performance ranges of a typical APD both at short and long wavelengths [7]. Table 6.4 illustrates the characteristics of a silicon APD as might be used with a short-wavelength application of moderate speed.

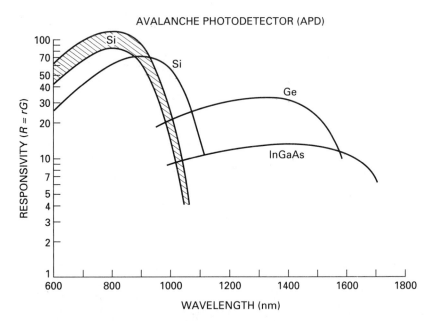

Figure 6.13   Avalanche photodetector responsivity characteristics.

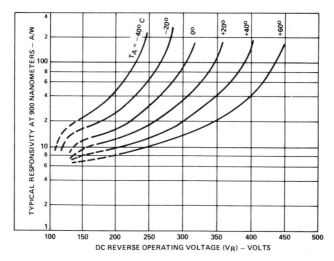

Figure 6.14 Typical responsivity at 900 nm vs. operating voltage for RCA C30954E avalanche photodiode. (From Ref. 12; courtesy of RCA.)

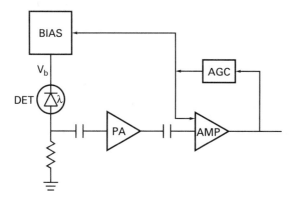

Figure 6.15 Receiver design for an avalanche photodetector illustrating automatic gain control to compensate for thermal drift and received power variations.

6.5 RECEIVERS

The output current of the photodetector is likely to be quite feeble. For a light output of 10 nW, for example, the detector output current may be as low as 5 nA for a PIN photodiode and 0.5 μA for an APD [8, p. 633].

It is the job of the receiver to amplify this current and recover either the digital or analog information from it. In the process it must not introduce large amounts of noise.

TABLE 6.4 ELECTRICAL CHARACTERISTICS OF RCA C30954E
AVALANCHE PHOTODIODE

Electrical characteristics at $T_A = 22°C$ at the dc reverse operating voltage V_R supplied with the device			C30954E Light Spot Diameter 0.25 mm (0.01 in)	
	Min.	Typ.	Max.	Units
Breakdown voltage, V_{BR}	300	375	475	V
Temperature coefficient of V_R for constant gain	—	2.2	—	V/°C
Gain	—	120	—	
Responsivity				
At 900 nm	65	75	—	A/W
At 1060 nm	30	36	—	A/W
At 1150 nm	4	5	—	A/W
Quantum efficiency				
At 900 nm	—	85	—	%
At 1060 nm	—	36	—	%
At 1150 nm	—	5	—	%
Total dark current, I_d	—	5×10^{-8}	1×10^{-7}	A
Noise current i_n				
$f = 10$ kHz,				
$\Delta f = 1.0$ Hz	—	1×10^{-12}	2×10^{-12}	$A/Hz^{1/2}$
Capacitance, C_d	—	2	4	pF
Series resistance	—	—	15	Ω
Rise time, t_r				
$R_L = 50$ Ω,				
$\lambda = 900$ nm,				
10% to 90% points	—	2	3	ns
Fall time				
$R_L = 50$ Ω,				
$\lambda = 900$ nm,				
90% to 10% points	—	2	3	ns

Source: Ref. 12; courtesy of RCA.

To reduce noise yet keep a proper dynamic range and signal bandwidth, optical receivers use different low-noise preamplifier designs tailored to the detector device and application. These designs will be discussed in the following paragraphs.

6.5.1 Receiver Operation

The basic block diagrams for digital and analog receivers are shown in Figs. 6.16 and 6.17, respectively.

In a digital receiver the received optical pulse train is converted to an electrical current pulse train by the photodetector and then to an amplified voltage pulse train by the following amplifier stages [7, 8]. The signal is amplified so that peak-to-peak

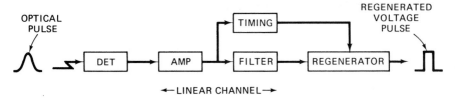

Figure 6.16   Digital receiver black diagram. (From Ref. 8, p. 629; copyright 1979 Bell Telephone Laboratories, Inc.; reprinted by permission.)

levels fit within a predetermined reference. The signal feeds both a timing recovery circuit as well as a digital signal regenerator.

The timing circuit is essentially a square-wave oscillator (referred to as a "clock") that runs at the approximate frequency of the clock on the transmit end. When it receives a signal it locks in frequency and phase to that signal, thus mirroring the clock on the transmit end. It uses this new clock pulse to determine where the midpoint of the incoming pulse should be so as to tell the regenerator when to sample it. The clock ensures that nearly all timing error (called jitter) on the incoming pulse is removed, by creating a new timing source for the leading and trailing edge of each pulse that comes from the regenerator.

The signal entering the regenerator is filtered to eliminate as much noise as possible and to shape the signal for proper presentation to the decision threshold logic. The decision threshold logic measures the height of the pulse at the instant in time that the timing circuit tells it to, presumably at the midpoint in the pulse period. If the pulse height is above a certain midpoint voltage (called the "threshold") it is decoded as a binary 1 and sent from the regenerator as a "high" state during a new pulse period as dictated by the timing circuit. Likewise a voltage lower than threshold will be decoded as a binary 0.

In the analog receiver the signal is detected and converted to a voltage waveform in the same fashion as with the digital receiver. It is then amplified to the level required for further processing, such as with a demodulator if the analog signal is on a carrier. Filtering is used to separate the desired signal channel from any others that might be present and to reduce noise as much as possible. (Remember that noise is a function of bandwidth, and filters reduce bandwidth.) The resulting waveform is fed to the demodulator where the signal is separated from the carrier.

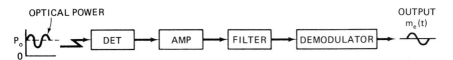

Figure 6.17   Analog receiver block diagram. (From Ref. 8, p. 630; copyright 1979 Bell Telephone Laboratories, Inc.; reprinted by permission.)

6.5.2 Fiber Optic Receiver Packaging

Generally the fiber-optic receiver is an entire printed-circuit card assembly that contains not only the detector, preamplifier, and regenerator or postamplifier stages, but also timing circuits, bias supplies, power regulators, alarm circuits, filters, and perhaps other signal conditioning circuits.

The heart of the assembly, and the most critical element from a performance standpoint, is the detector-preamplifier assembly. This assembly must be packaged as one unit to provide the optimum in low noise performance, because coupling capacitances and inductances between the detector and the preamplifier first stage transistor must be kept at a minimum. Also the package must be fully enclosed by metal to provide proper shielding from outside interference. Because the detector is inside, optical coupling must be with an integral optical connector or a factory-installed fiber pigtail with external connector.

Some of the more common packaging designs are illustrated in Figs. 6.18, 6.19, and 6.20. Figure 6.18 illustrates an open window design useful for only slow-speed low-performance designs because the detector is required to be large for light capture through the glass window. The design in Fig. 6.19 contains an integral connector with a large core fiber that is factory coupled to the detector for minimal coupling loss. Figure 6.20 illustrates the integral fiber pigtail design most often used in high-speed, long-distance application. The fiber pigtail is generally a small core (50-μm or so) multimode fiber so that it couples well to single-mode fiber with minimal loss yet permits the connector size to remain small.

Figure 6.18   Integrated PIN photodiode and preamplifier. (Courtesy of RCA.)

Figure 6.19   Integrated APD and preamplifier. (Courtesy of RCA.)

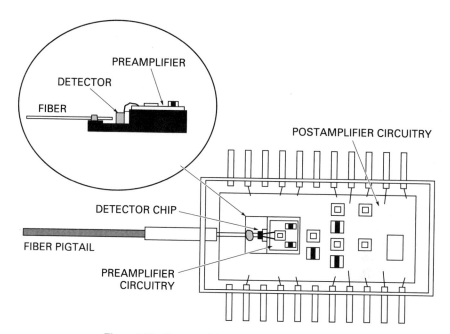

Figure 6.20   Conceptual drawing of a packaged receiver.

6.5.3 Preamplifier Design Choice

The selection of preamplifier design depends on the detector chosen (PIN or APD), the bandwidth required, the dynamic range necessary, and such practical considerations as cost.

The key element in preamplifier design is the amplifying device. The choice is generally between a bipolar transistor and an FET transitor. FETs perform better but are more expensive. The GaAs FET is generally the best performing device.

The choice of design also depends on the detector. If an APD is chosen the noise performance of the receiver is more dependent on the APD quantum noise than on the amplifier noise, and therefore preamplifier design becomes less critical.

Figure 6.21 illustrates the three receiver preamplifier designs most often used with fiber optics [7]. The voltage amplifier is simplest and generally the lowest in cost, but is the poorer performer. Signal voltage is produced by the photocurrent flowing through the input resistor R_L. The amplifier simply amplifies this voltage. The limitation is that to make the preamplifier low noise, R_L must be made a very high value to reduce thermal noise current. Because the frequency response (bandwidth) of the amplifier is inversely proportional to R_L multiplied by detector and amplifier capacitance, increasing R_L decreases system bandwidth. This design is therefore limited to relatively short, slow-speed systems with moderate performance requirements.

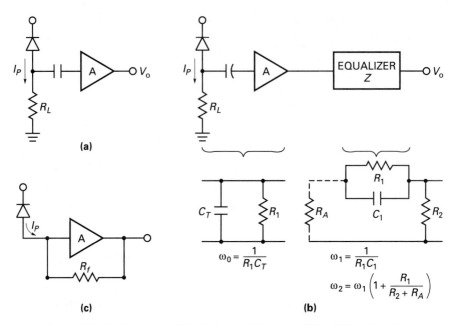

Figure 6.21 Receiver preamplifier design. (a) Voltage amplifier, (b) high-impedance integrating, and (c) transimpedance.

The high-impedance integrating design is similar to the voltage amplifier, but it overcomes the noise and bandwidth problem with a compensation circuit in the following stages of amplification. It essentially uses a high impedance at R_L for low noise, and permits the bandwidth of the first stage to be reduced. The following stages of amplification contain an equalizer. It is a differentiator network that increases gain with frequency, thus compensating for the reduced bandwidth of the first stage. The disadvantage of this circuit is the amount of dynamic range that is consumed by equalization, makes the overall dynamic range performance for the receiver much less (about 10 times less) than other designs such as the transimpedance. Dynamic range is the ability to handle small and large signals. Because this design is one of the lowest noise, it is often used with fixed point-to-point long-distance transmission systems where signal levels remain relatively fixed.

One of the most popular designs is the transimpedance. It uses feedback to reduce apparent input impedance. This permits fast response time owing to the low effective input resistive-capacitive time constant that limits bandwidth in other designs. The signal voltage is still developed across R_L, but because it can now be made large, signal level is high while thermal noise is low. Dynamic range is also large. It seems only limited by the performance of transistors and components that make it up. Transimpedance amplifiers are used in high-speed, long-haul systems where low noise and high bandwidth is a requirement, as well as in shorter slower-speed LANs where dynamic range is critical.

Figure 6.22 illustrates the range of sensitivity that can be expected of the various combinations of preamplifier design and photodetector [7]. The minimum required detected power (MRP) is the amount of optical power required at the detector to achieve 10^{-9} BER or better for a digital signal. Note that as bit rate increases, the bandwidth required increases, thus increasing receiver noise. The result is that more optical power is required at higher bit rates to overcome the increased noise.

6.5.4 Receiver Design Examples

The design for a medium bandwidth (up to 50 Mb/s) medium distance integrated detector with transimpedance amplifier is illustrated in Fig. 6.23. It requires a minimum of 2.0 μW of light for a 10 dB SNR [9]. Figure 6.24 illustrates its use in a complete receiver. Note that this receiver uses no regenerator, and therefore is susceptible to pulse jitter and distortion if used in a demanding application. It depends on the differential operation as well as saturation of the final amplifier stage to square up both the positive- and negative-going pulses.

Figure 6.23 illustrates an early moderate speed transimpedance design developed by Motorola [10]. Although it uses discrete components and is not as integrated and advanced as current designs, it illustrates some of the principles of the trans-

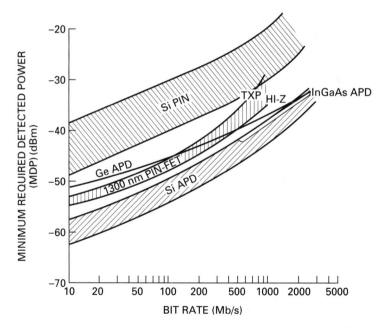

Figure 6.22   Typical receiver sensitivity representative of various designs for 10^{-9} BER. (TXP = transimpedance; HI-Z = high impedance integrating. (From Ref. 7; copyright 1990 Prentice Hall, Inc.)

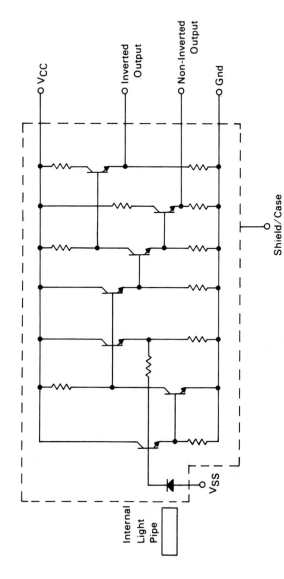

Figure 6.23 MFOD403F integrated detector/preamplifier equivalent schematic. (From Ref. 9; copyright 1979 Motorola, Inc.)

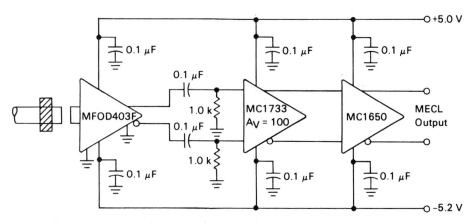

Figure 6.24   Ac-coupled 40-Mb/s MEDL output fiber-optic receiver. (From Ref. 9; copyright 1979 Motorola, Inc.)

impedance design. It represents only the first stages of a receiver and does not include regeneration.

The receiver uses an MFOD100 PIN photodiode as an optical detector. The detector diode responds linearly to the optical input over several decades of dynamic range.

The PIN detector output current is converted to voltage by integrated circuit U1 (operational amplifier LF357). The minimum photocurrent required to drive U1 is 250 nA.

Receiver dynamic range is extended with diode D2 to prevent U1 from saturating at large optical power inputs.

Integrated circuit U2 acts as a voltage comparator. Its worst-case sensitivity of 50 mV determines the size of the pulse required out of U1. U2 detects, inverts, and provides standard TTL logic levels to the output.

CMOS compatible operation is available when integrated circuit U3 is wired into the printed-circuit board. This IC is an open-collector TTL quad, two-input NAND-gate device. Jumper wire J1 must be connected from U3 output to the receiver output terminal.

Figure 6.25 illustrates the schematic for an integrated amplifier/detector/fiber pigtail receiver preamplifier manufactured by AT&T [11]. The performance characteristics for this device are given in Table 6.5. The receiver uses a high-performance InGaAs PIN photodiode followed by a low-noise GaAs FET transimpedance front-end preamplifier. The design has a capability of 12 Mb/s and employs an external equalizer to boost performance to 44.7 Mb/s.

The preamplifier package features a 70-μm core multimode graded-index fiber pigtail with a biconical connector on the other end. The failure rate is estimated at less than 105 service hours.

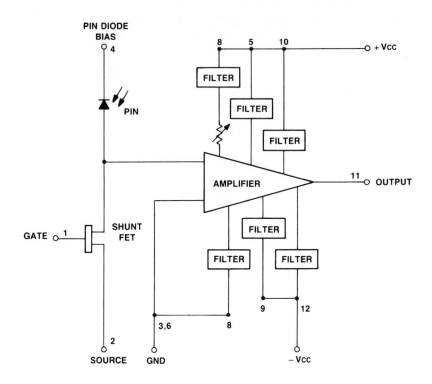

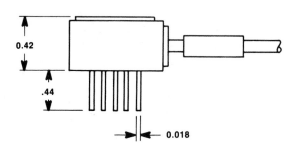

Figure 6.25 Integrated lightwave detector and preamplifier. (From Ref. 11; courtesy of AT&T.)

TABLE 6.5 PERFORMANCE OF AT&T INTEGRATED PREAMPLIFIER
(AT NOMINAL SUPPLY VOLTAGES AND AMBIENT OF 23°C)

Electrical characteristics					
Parameter	Symbol	Min.	Typ.	Max.	Unit
Transimpedance at 1 MHz	Z				
1300 D		—	500	—	kΩ
1300 J		—	125	—	kΩ
DC power-supply voltages					
	V_{BIAS}	5	10	20	V
	$+V_{CC}$	—	5.0	—	V
	$-V_{CC}$	—	-5.2	—	V
Power-supply current drain					
	I(BIAS)	—	—	1	mA
	I(V_{CC})	—	—	60	mA
	I($-V_{CC}$)	—	—	5	mA
Input GaAs FET					
Transconductance		—	30	35	40	mS
Gate leakage current		—	—	—	3	nA
Saturated drain current	I_{DSS}	20	40	60	mA
Shunt GaAs FET					
Pinch-off Vgs		—	-3.7	-3.0	- 2.5	V
Pinch-off I_D		—	—	—	85	nA
On resistance R_{DS}		—	—	—	40	Ω

Optical characteristics					
Parameter	Symbol	Min.	Typ.	Max.	Unit
Optical wavelength for rated sensitivity	λ	1.25	1.3	1.35	μm
Useful range of optical wavelength	λU	1.1	—	1.6	μm
Minimum optical input power at pigtail					
10^{-7} BER	P_{IL}				
1300 D at 12.6 Mb/s		-50.5	-52	-53	dBm
1300 J at 44.7 Mb/s		-44.6	-46	-47	dBm
Maximum optical input power at pigtail					
10^{-7} BER (with shunt FET turned on)	P_{IH}				
1300 D at 12.6 Mb/S		-12	—	—	dBm
1300 J at 44.7 Mb/s		-12	—	—	dBm
Pin photodetector					
Dark current	I_D	5	10	20	nA
Capacitance		—	0.55	0.60	0.65	pF
Responsivity		—	0.5	0.65	0.8	A/W

REFERENCES

1. "Introduction to Fiber Optics and AMP Fiber-Optic Products," AMP HB 5444, AMP Incorporated, n.d.
2. R. Hoss, *Fiber Optic Communications Design Handbook* (Englewood Cliffs, N.J.: Prentice Hall, Inc. 1990), pp. 141–180.
3. Joseph Zucker, "Choose Detectors for Their Differences to Suit Different Fiber-Optic Systems," *Electronic Design,* Vol. 28, No. 9, April 26, 1980.
4. Tingye Li, "Optical Transmission Research Moves Ahead," *Bell Laboratories RECORD,* Sept. 1975, pp. 333–340.
5. "Optical Communications Products," RCA OPT-115.
6. "MFOD104F Fiber Optics PIN Photo Diode," Motorola Advance Information, 1979.
7. R. Hoss, "Receiver and Detector Specifications," in *Fiber Optic Communications Design Handbook,* (Englewood Cliffs, N.J.: Prentice Hall, Inc. 1990), pp. 70–88.
8. Stewart D. Personick, "Receiver Design," in *Optical Fiber Telecommunications,* ed. Stewart E. Miller and Alan G. Chynoweth (New York: Academic Press, Inc., 1979).
9. "MFOD403F Fiber Optics Integrated Detector Preamplifier," Motorola Semiconductors Product Preview, 1979.
10. "Basic Experimental Fiber Optic Systems," Motorola Semiconductors Advance Information, April 1978.
11. AT&T datasheet, "1300D/J Lightwave receivers," AT&T Technologies, July 1985.
12. "Photodiode Developmental Types," RCA Electro Optics and Devices, Oct. 1979.

7

System Design
and Architecture

Although this book is not intended to be an engineering guide to system design, some understanding of the process provides a good foundation for anyone working with any aspect of fiber optics, be it planning, network design, installation, or maintenance. A summary of the design process is given here; a complete design handbook can be found with reference [1].

The proper design of a fiber-optics system, or any communications system for that matter, follows the flow illustrated in Fig. 7.1. The process is iterative in that candidate designs are tried and analyzed and adjusted until the intended operating requirements are met.

The designer begins by developing an end-to-end system requirements specification which includes signal performance, reliability, physical, environmental, and cost requirements. Candidate architectures (functional block diagrams) are then developed and components selected or specified to fit those architectures, usually based on supplier's data. The end-to-end performance is then analyzed to see if the design works as specified. Components and architecture are adjusted to achieve or optimize performance.

Once signal performance is achieved, system reliability is analyzed, and redundancy or alternative components are added if the requirement is not achieved.

Finally the system cost is determined to see if it achieved the cost constraints. Sometimes when cost is a major consideration, a cost versus performance model is developed, before an in-depth signal performance analysis is done, in order to select the best architectural candidate.

The remainder of this chapter will concentrate on system performance analysis and system architecture.

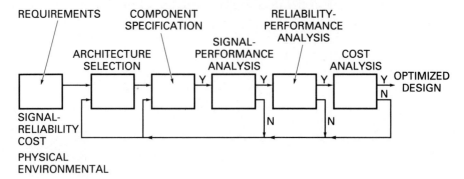

Figure 7.1 Communications systems design methodology. Y = yes; N = no.

7.1 SYSTEM PERFORMANCE ANALYSIS

To understand how to use fiber optics to meet a particular communications design problem, the complete system must be considered. A typical system block diagram for a point-to-point system is illustrated in Fig. 7.2. The system not only includes the electro-optics and fiber, but the modulation, multiplexing, and encoding.

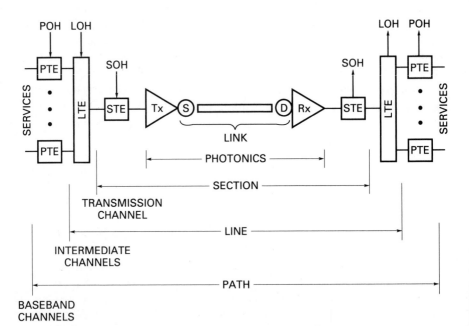

Figure 7.2 Generalized system functional block diagram for point-to-point fiber-optics transmission system with multiplexed analog or digital channels. Tx = transmitter; Rx = receiver; POH = path layer overhead; LOH = line layer overhead; SOH = section layer overhead.

The functional blocks within the system model are defined here by taking the liberty of borrowing and generalizing some of the terminology from the SONET standard [2]. (Note that the definition of the layers as presented in the standard are more specific to the SONET multiplexing approach, but the function is similar to that presented here.)

1. *Photonics* (PHOT). The fiber optic transmitter, receiver, source, detector, fiber, connectors, couplers, and other optical components.
2. *Section* (STE). The interface and signal conditioning electronics for the optical transmitter and receiver. It is where the signal is finally conditioned in the proper format for the optical transmitter.
3. *Line termination equipment* (LTE). Generally the final multiplexer that combines all signal channels into one, as well as any overhead (LOH).
4. *Path termination equipment* (PTE). The equipment that converts the various originating signals, analog or digital, into a channelized format that is compatible with the line termination multiplexer. It includes the modulation electronics as well as any lower-level multiplexing.

For simplicity let us bundle all the path and line termination equipment under one category called "signal conditioning." In today's product environment much of the signal conditioning equipment is somewhat standardized subsystems that have been developed for various classes of networks. Such classifications include different but standard types of equipment for long-distance trunking or equipment for LANs. The remainder of this chapter is devoted to the different network classifications for which standard equipment design are established, and the general design of those subsystems. This section concentrates on the design approach for the fiber-optics subsystem that we will call the photonics or fiber optics link.

7.1.1 Systems Design Procedure

There are two basic stages in the design of a fiber optics system.

Stage 1.   Design the systems architecture and select/design modulators and multiplexers with the goal of determining end-to-end photonics performance requirements.

Stage 2.   Select fiber-optic components and perform link budget analysis to determine if the photonics design meets the requirements of stage 1.

The design approach is iterative so that if the initial selection of architecture and components in either stage does not work, then the designer modifies the selection and reattempts the analysis until the design is optimized.

7.1.2 Stage 1—System Architecture

Step 1A—Requirements.   Define the system end-to-end design requirements including (1) location of system terminals, end-to-end distances, and cable routing; (2) signal performance (SNR or BER and bandwidth or bit rate); (3) environmental requirements; (4) physical requirements; (5) system availability or reliability; and (6) system cost constraints. These will be used as the criteria to select candidate designs and components as well as the criteria by which the design will be evaluated.

Step 1B—System Architecture.   The basic functional block diagram of the entire system including the photonics should be detailed (as in Fig. 7.1). Select the candidate modulation, and multiplexing approach to be used (the signal processing subsystem). The remainder of this chapter, as well as reference to Chapter 3 (Section 3.4) will assist in this decision process. There is no need to select the fiber technology at this time until the requirements of the signal processing subsystem are determined. These will dictate photonics end-to-end performance.

Step 1C—Select Signal Processing Components.   The signal processing equipment (modulators and multiplexers) should be selected, generally from manufacturers data sheets. Selection criteria is based on the set of requirements defined in Step 1A. An alternative for custom applications is to design, or specify the performance characteristics of, a unique modulator or multiplexer. This is beyond the scope of this book; however, reference [1] provides an in-depth guide to these design procedures.

Step 1D—Define Photonics Performance.   The output signal of the final line multiplexer (transmit end LTE) and the input requirements of the line demultiplexer (receive end LTE) define the signal interface and performance parameters of the photonics section. As a result of modulation and multiplexing, the photonics performance will generally be of a completely different signal format and bandwidth than that of the original signal or signals. For example, voice channels may be pulse-code modulated and time-division multiplexed to form a digital pulse stream for transmission, which requires orders of magnitude more bandwidth and orders of magnitude less SNR than each voice channel.

7.1.3 Stage 2—Photonics Design

Step 2A—Technology Selection.   When the multiplexer is selected and purchased, the signal input and output interfaces generally follow standard data rates, formats, and signal levels; for example, 1.544 Mb/s DS1 input and 44.736 Mb/s DS3 output. As a result of standardization, there will be various photonics products available to select from that are compatible with the multiplexers. The objective of this step is to select the fiber optics technology, from among those available products, that will best achieve the signal performance (bit rate and BER, or bandwidth and SNR) over the required end-to-end transmission distance, with the least number of repeaters.

Technology selection at this point in the process is not product selection but is aimed at determining such things as whether short or long wavelength should be used, and whether the fiber should be single mode or multimode. Selection is generally made with the aid of product literature, however. Fig. 7.3 can also be used as a guide in selecting the proper technology based on distance between terminals or repeaters, and transmission bit rate.

Step 2B—Select Components and Determine Span Distance.   Once the type of technology to be used is identified, select the specific components and terminal equipment products that appear to be the best candidates. Often the final multiplexer stage is sold as an integral part of the optical selection and photonics equipment, that is, the line, section, and photonics termination equipment are integrated. For example, a 565-Mb/s fiber terminal may include a 12 × DS-3 multiplexer as a part of the fiber terminal. In this case LTE and photonics technology are selected together.

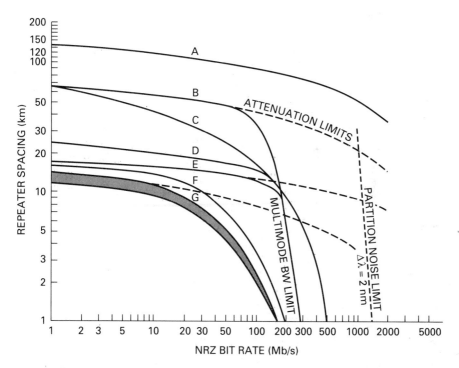

Figure 7.3   Length bandwidth capability of fiber product technology. Length bandwidth capability for various component mix: A = 1300 or 1550 nm, SM, ILD, PINFET det., 0.3 dB/Km SM fiber; B = 1300 nm, MM ILD, PINFET det., 0.7 dB/Km graded-index fiber; C = 1300 nm, MM ILD, PINFET det., 1.8 dB/Km partially graded fiber; D = 820 nm, MM ILD, APD det., 3.0 dB/Km graded-index fiber; E = 1300 nm, SM LED, PINFET det., 0.4 dB/Km SM fiber; F = 820 nm, MM LED, APD det., 3.0 dB/Km partially graded fiber; G = 820 nm, MM LED, PIN det., 3.0 dB/Km partially graded fiber. (From Ref. 1; copyright Prentice Hall, 1990.)

At this point determine from the product literature (or design aids such as Fig. 7.3) approximately how many repeaters may be required and the approximate span distance of each optical section.

Step 2C—Link Power Budget.   Fig. 7.4 illustrates elements of a fiber optics link with the points of optical power injection and loss identified. This link is a somewhat generalized reference model, but contains the elements of both a point-to-point system as well as a LAN. It is used to calculate loss between any transmitter and receiver span in a system.

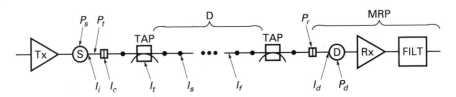

Figure 7.4   Fiber optics photonics design reference model with power budget parameters identified.

Draw such a diagram for your system with the selected components. Obtain and record the typical, best-case and worst-case performance specifications for each component where available from product literature or test data.

The objective in this step will be to determine whether there is enough power at the receiver, after system losses and operating margins are considered, for the system to work as specified. If there is more power than is needed it is called excess power. The process is called a link power budget.

The relationship for the link power budget is

$$Excess\ Pwr = source\ Pwr - losses - margin - Rcvr\ sensitivity$$

Proper design results in an excess power of zero or greater. In terms of the parameters in Fig. 7.4:

$$Pe = Pt - (Lc + Lt + Lf + Ls) - M - MRP$$

where

Pe = excess power in dB (should be 0 dB or greater)

Pt = source power coupled into the fiber in dBm (1 mW)

Lc = connector losses = lc (dB/conn) × no. connectors

Lf = total fiber loss = lf (dB/km) × distance (km)

Ls = total splice loss = ls (dB/splice) × no. splices

M = operating margin, which consists of (1) a safety margin of the designer's choice (typically 3 dB) plus (2) margin for transmission noise, bandwidth limiting, and signal distortion that were not considered by the receiver or transmitter specification

MRP = receiver sensitivity expressed as the minimum required optical power in dB (1 mW), at the connector to the optical receiver, necessary to provide desired performance (SNR or BER).

Note that the insertion losses of the connectors, splices, couplers, and fiber are given in a somewhat generalized form in the preceding equation, which principally represents a point-to-point system. Fig. 7.5 illustrates the loss relationships that might result from other fiber optic systems topologies.

Step 3D—System Performance Test.   System performance needs to be calculated for the typical case as well as for the case in which component tolerances and environmental conditions are at their worst. This will determine whether the operating margin is adequate under all conditions.

If the information is available, it is also a good idea to calculate best-case performance to see whether the system will deliver too much power to the receiver under certain conditions and saturate it. Generally the receiver literature will specify maximum power.

The solution for saturation conditions is generally to add an attenuator to the jumper cable at the receiver if it occurs. Alternatively, if the receiver is close to saturation while adequate margin is achieved under worst-case conditions, then longer distances between repeaters (if any), or a lower performance and perhaps lower-cost fiber technology should be considered.

7.1.4 Practical Considerations in Performing the Link Budget Analysis

Reference [1, pp. 263–311] provides an in-depth procedure for performing system power and bandwidth performance analysis. Some of the practical considerations provided in that reference will be summarized here.

7.1.4.1 Statistical Behavior of Components

When components are ordered from a supplier the performance is generally specified to fall within a certain acceptable range, usually given as a typical and worst-case performance. These values are given to describe the quality-control limits and expected performance (tolerances) for devices that fall within a range acceptable for shipment. Where a specific device performs within this range is unknown until received, but if a history is kept of performance measurements over a meaningful sample of devices manufactured (known as a histogram), then the probability that a device will perform to a specific level can be determined.

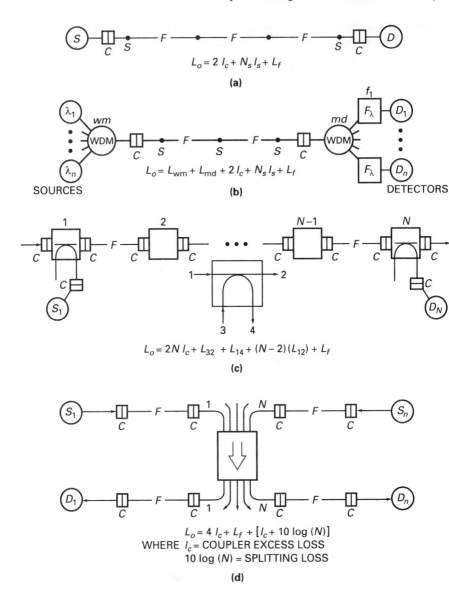

Figure 7.5 Optical loss-budget relationships for various architecture. (a) Point-to-point simplex link, (b) point-to-point wavelength multiplexed, (c) T bus, and (d) star bus.

An example of such a histogram was given in Chapter 5 illustrating how when connectors and splices join their tolerances interact in somewhat of a random fashion that, when plotted, fall within a certain predictable range.

The shape of the resultant histogram can be described statistically if related to a "normal" distribution curve. All component histograms do not precisely follow a "normal" distribution, but they provide good approximations. Typical component performance can thus be defined in terms of mean performance. The amount that components vary from the mean from device to device can be described by the term *standard deviation*.

A normal distribution curve is illustrated in Fig. 7.6. The mean relates to the value that 50% of the components would be expected to perform at. One standard deviation relates to the amount of deviation from the mean that 84.8% of the components will fall within. The maximum or minimum value specified for a component is most commonly 3 SDs from the mean. This is the value that 99.87% of the components can be expected to be better than.

When component performance is specified statistically, generally the mean and the 1 SD (1 σ) values are given. The typical, best, and worst case performance can be derived from these measurements. Typical performance is simply the mean. Worst- and best-case performance is generally defined as the situation that occurs only 0.1% of the time or 3 SDs from the mean (although some manufacturers use 2). The worst-case value, therefore, can be found by adding the mean value to 3 times the SD value.

When components are connected in series, their mean values add together di-

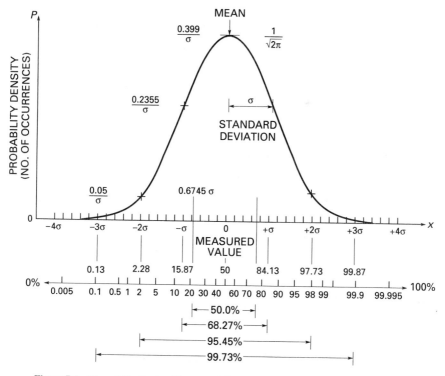

Figure 7.6   Normal distribution. The normal distribution curve is marked to indicate mean and SD relationships, and the percentage of measured values that can be expected to fall between designated points on the curve. (From Ref. 1; copyright Prentice Hall, 1990.)

rectly, but the SD values do not. SDs add as the square root of the sum of the squares. For example, to find total loss for a group of optical components in series, one would add the mean losses together and add to it 3 times the square root of the sum of the squares of the 1=SD values. For a number (N) of like components of individual loss (l), the total worst-case loss (L) calculation can be simplified to:

$$L(\text{dB})wc = N \; l(\text{dB}) \text{ mean} + 3\sqrt{N} \; 1(\text{dB}) \text{ SD}$$

Where the manufacturer only provides typical and worst-case values, a rough approximation to mean and standard deviation can be made by assuming that the worst-case values are measured at the 3-SD point. The 1-SD value can be derived from subtracting the worst-case value from the typical and dividing by 3. Using this assumption the total worst case loss (L) of a number (N) of like components of individual loss (l) acting in series can be simplified to

$$L(\text{dB})wc = N \; lf(\text{dB})\text{typ} + \sqrt{N} \; [1(\text{db})wc - 1(\text{dB})\text{typ}]$$

7.1.4.2 Transmitted Power (Pt)

Generally the manufacturer of the source or transmitter provides typical, minimum, and maximum values of coupled power. The minimum values may include temperature and lifetime considerations as well. Note also that the coupled power value may be referenced before or after the attached connector. If afterward, do not count the connector loss in Lc computations. Use the minimum and maximum values in the worst- and best-case analysis, respectively.

7.1.4.3 Connector Loss (Lc)

Connector loss may be different at the detector than at the source. Also note that if source or receiver coupled power specifications include the attached connectors, do not double count them in the analysis.

For the typical-case analysis, use mean connector loss. For best-case, use zero loss. For worst-case analysis, use worst-case loss if there are only two connectors in line. If more connectors are present in series, it is best to use statistical measurement data if available or use the approximation given in Section 7.1.3.1.

If statistical data is presented by the manufacturer the total worst case connector loss (Lc) for a number (Nc) of connectors of individual loss (lc), acting in series, is estimated as

$$Lc(\text{dB})wc = Nc \; lc(\text{dB})\text{mean} + 3 \sqrt{Nc} \; lc(\text{dB})\text{SD}$$

7.1.4.4 Fiber Loss (Lf)

Fiber loss (attenuation) is given by the supplier in decibel per kilometer of cable. The supplier gets this information by measuring the attenuation of the cable on the reel and dividing it by the reel length. The system designer, however, generally does not have this information but simply the supplier's specification of nominal and worst-case attenuation and sometimes the worst-case attenuation increase with temperature.

In practice, when the cable is delivered, the worst-case measurements are generally only observed on a few fibers in a few reels of delivered cable. Assuming worst-case performance for all fibers, when designing the system, will generally lead to overdesign.

If only the typical and worst-case measurements are given, it is acceptable to perform the analysis both assuming typical then worst-case measurements. If the difference between worst case and typical performance results in a major design change to make the worst-case conditions work, then a different approach should be tried. You can gain a more realistic worst-case result by using the approximation given in Section 7.1.3.1 by substituting fiber distance in kilometers (Df) for N and unit fiber unit loss in decibels per kilometers for lf.

If the statistical performance parameters are given by the supplier then worst-case end-to-end loss can be calculated as

$$Lf(\text{dB})wc = D(\text{km}) \; lf(\text{dB/km})\text{mean} + 3\sqrt{D} \; lf(\text{dB/km})\text{SD}$$

7.1.4.5 Splice Loss (*Ls*)

Splicing loss is difficult to determine since it depends on the splicing equipment, the fiber tolerances between reels, and the quality of workmanship. Some values are given in Chapter 5, but it is best to rely on past experience. The manufacturer of the splice equipment or a splicing subcontractor should have this information. It is best to calculate splice loss based on statistical measurements. If not available then typical and worst case can be used, but again remember that it is unlikely all splices in a fiber will come out as worst case, so use the approximation in Section 7.1.3.1.

The sum of the splice losses within a link, containing Ns splices, is as follows:

$$Ls(\text{dB})wc = Ns \; ls(\text{dB})\text{mean} + 3\sqrt{Ns} \; ls(\text{dB})\text{SD}$$

Sometimes the number of splices are not known when repeater distance is an unknown. In this case simply divide per splice loss by average reel length (or estimated distance between splices) and add the resultant per kilometer loss to the per kilometer fiber loss.

7.1.4.6 Receiver Sensitivity (*MRP*)

The minimum required received power (*MRP*) is specified at a specific signal bandwidth or bit rate, and for a specific signal performance. For example, in most fiber equipment a BER of 10^{-9} is considered the minimum performance level and 10^{-11} the typical. Depending on the product, the received power measurement is specified either as power at the connector, or as the power coupled into the pigtail or detector aperture after the connector.

When the receiver is selected as a part of a matched pair to the transmitter (which is usually the case) the receiver specification accounts for any signal distortion and noise caused by the transmitter (including extinction ratio). In cases in which the receiver does not account for these effects, they must be accounted for in the margin calculation.

Generally a maximum received power is given. This is the point at which the receiver begins to saturate and no longer works to specification. This value is used in the best-case analysis to determine whether the receiver will saturate under best-component performance conditions.

7.1.4.7 Margin (*M*)

Margin is added in the design as a safety factor, for degradation with time and temperature, as well as to account for any known optical noise and signal distortion that will affect receiver performance.

Operating Margin.   Operating margin is simply a safety factor to account for effects that the analysis did not include. Typically a 3-dB margin is used (received signal level double that needed) depending on the thoroughness of the analysis as well as other factors.

Bandlimiting Margin.   When a signal passes through the optical components the bandwidth of the signal is reduced somewhat by the rise and fall times of the source and the detector as well as by the dispersion of the fiber.

If the signal is a digital pulse stream, the result is rounded pulses that have "tails" that extend into adjacent pulses. This is referred to as pulse interference, which results in errors in decoding, that is, an increased BER. This can be compensated for somewhat by increasing optical power, thus adding a margin to the design.

When purchasing a matched receiver/transmitter pair, the effect of the response time of the electro-optical components is accounted for in the receiver sensitivity value. The effect of the fiber is not, however. The receiver manufacturer should provide a curve or derating factor to compute the margin that must be added. This factor is generally calculated on the basis of signal bit rate vs. fiber bandwidth over the span length [1, p. 291]. Total fiber bandwidth is computed most easily in terms of dispersion (*td*) and then converted to bandwidth [1, p. 278]:

$$td(\text{total}) = \sqrt{td(\text{multimode})^2 + td(\text{material})^2}$$

where (see Chapter 5)

$$td(ns)\text{multimode} = tdw(\text{ns/km}) \times D^8 \text{ (km)}$$

$$td(ns)\text{material} = tdm(\text{ps/nm/km}) \times \Delta\lambda \text{ (nm)} \times D(\text{km}) \times 10^{-3}$$

The equivalent electrical (post detection) and optical fiber bandwidths are approximated as [1, p. 280]:

$$BWf(\text{MHz})\text{elect} = 312 \,/\, td(\text{ns})\text{total}$$

$$BWf(\text{MHz})\text{opt} = 441 \,/\, td(\text{ns})\text{total}$$

With analog transmission, bandwidth limitations in the fiber, source, and detector reduce the amplitude of the signal at higher frequencies. As a result added power is needed at the higher frequencies. The amount of power margin that should be added for a transmitted signal having a signal frequency or bandwidth (ft) is [1, p. 282]

$$M = -X \log [\ 1 - BW(\text{link})^2 \,/\, \text{ft}^2]$$

where X is 5 for most APD receivers and 7.5 for most PIN/FET receivers.

Link bandwidth, $BW(\text{link})$, is determined most easily from summing the rise-times of the source (trs), the detector (trd), and the fiber (trf), then converting to bandwidth [1, p. 278]:

$$tr(\text{link}) = 1.1\sqrt{trs^2 + trf^2 + trd^2}$$

$$trf = 1.087 \, td(\text{total})$$

$$BW(\text{link}) = 0.35/tr(\text{link})$$

Modal Noise Margin.   Noise can be created within the source and the fiber from the interaction of various modes of light that are present. This noise can result from "beat" noise in wide spectrum sources, speckle pattern noise in multimode fiber driven by laser sources, or partition noise in single-mode systems.

Speckle noise is generally of concern only in multimode fiber systems that use lasers. Its effect is most severely seen when trying to transmit direct intensity-modulated analog information (such as video signals) on multimode fiber. It is caused by reflections from fiber joints (connectors and splices) and selective coupling of modes at these joints. The effect is reduced by using only digital transmission with laser/multimode fiber systems, proper transmitter design, and use of LEDs or lasers with many spectral lines to cancel out the effect.

Of the three noise sources, the one that generally is most often of concern is partition noise in very long, high-speed single-mode systems. The required power margin is computed as follows [1, p. 297]:

$$M = -5 \log[\ 1 - (SNRt \,/SNRn)^2]$$

where:

$$SNRt = \text{peak signal level} \,/\, 2 \text{ rms noise level}$$

$$= 6 \text{ for a } BER \text{ of } 10^{-9}$$

$$SNRn = \frac{\sqrt{2}}{[\pi\,(BR)\,D\,\sigma_\lambda\,tdm\,10^{-12}]^2}$$

where

$$BR = \text{bit rate (bits per second)}$$

$$D = \text{fiber length in km}$$

$$\sigma_\lambda = \text{rms source spectral width in nanometers}$$

$$tdm = \text{material dispersion in ps/nm/km}$$

Extinction Ratio. Extinction ratio (ER) is the fraction of total optical power emitted by the source that is used by a binary signal. It is related as

$$ER = \text{power in low state / power in high state}$$

It is common in laser diodes to have a significant extinction ratio because of the existence of a lasing threshold where the device is biased for the binary 0 or low state condition.

As noted in Chapter 6, optical power on the detector creates quantum noise. The receiver sensitivity specification (*MRP*) accounts for quantum noise under the assumption that all transmitted source power is signal. When some of the power is not signal then this added noise must be accounted for. When the transmitter and receiver are purchased as a pair this noise associated with the transmitter extinction ratio is generally accounted for in the receiver specifications. When the receiver is not purchased as a mated pair with the transmitter, however, power margin to compensate for the added noise must be added into the design. For extinction ratios of 0.05 and 0.1, margins of roughly 0.5 dB and 1 dB, respectively, are required [1, p. 291].

7.2 SYSTEM ARCHITECTURE

With the ever increasing integration of voice, video and data over common (mostly digital) transmission medium, the classical division of transmission product into voice, video, and data is diminishing. Product and technology today can be more appropriately differentiated by the scale and architecture of the transmission system. The division by wide-area, metropolitan-area, and local-area networks, better differentiates technological product differences.

7.2.1 Wide-Area Network (WAN)

WAN implies long haul trunking between channel concentration points, telephone offices, switching centers, or channel drop and insert points. The fiber-optic trunking networks that carry telephone communications between cities and around the world are examples of WANs.

7.2.2 Metropolitan-Area Network (MAN)

A MAN is a network that transports signals within a traffic concentration area, such as within a city, between buildings, between LANs, or for local distribution of WAN traffic. These are generally considered self-contained networks, with dedicated and switched bandwidth, often with gateways to WANs. Examples include telephone local loop, interexchange carrier alternate access business networks, CATV networks, and private business networks linking buildings within a limited geographical area.

7.2.3 LAN

LANs are multiple-access transmission facilities within an office, a building, or a campus of buildings within a small geographic area. They include private computer networks linking mainframes, storage devices, data terminals, and other peripherals. Examples include Ethernet, IBM Token Ring, and the Fiber Distributed Data Interface (FDDI).

As LANs extend into a network of LAN interconnections within a campus of buildings the term *campus-area network* (CAN) is sometimes used. The communication system that permits the internetworking of LANs within the wide or metropolitan area is often termed *LAN inter-network* (LIN).

The remainder of this chapter deals with the more common architectures and standards used in the application of fiber optics to the transmission of voice, video, and data in the LAN, MAN, and WAN.

7.3 WAN ARCHITECTURE

WANs are designed to trunk large quantities of data, voice, and video traffic between concentration centers or MANs. In public telecommunications networks a WAN is typically a mesh-type network where switching points provide alternate paths for calls in the event of a path outage. Route redundancy that a mesh network offers is required for high speed fiber optics trunks, because cable is susceptible to accidental cuts on a regular basis.

In public switched telephone networks, traffic is routed from the user to a Local Exchange Carriers Central Office (LEC/CO) and then to the long distance carriers point-of-presence (POP). POP to POP communications occurs over one path through the network. In the event of an outage on this path the traffic could be rerouted along an alternate path by the voice switch or digital cross connect (DACS) within the POP.

7.3.1 WAN Fiber Transmission

Figure 7.7 illustrates the functional block diagram of a fiber optic trunking system that might serve the telco network described above. At the LEC/CO individual subscriber voice or data channels are received by a switch or DACS, which multiplexes them onto higher speed trunk groups (generally DS-1 or DS-3) and passes them, along with signaling information, to the long-distance carrier's POP.

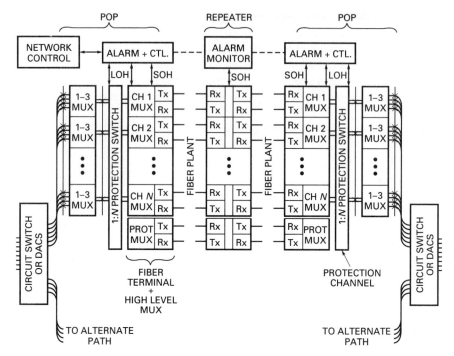

Figure 7.7   Fiber-optic trunking system for telecommunications application. (From Ref. 1; copyright Prentice Hall, 1990.)

Within the POP, the channels will enter the long distance carriers switch or DACS. This equipment connects each channel to a trunk group that is destined for the appropriate geographical destination. The channels bound for a particular long-haul trunk group will be multiplexed to a DS-3 rate (if not achieved in the DACS) and connected through a protection switch to the fiber-optic terminal equipment.

The fiber-optic terminal equipment contains high-speed multiplexers that will combine the channels further into groups of typically 12, 24, 48, or 96 DS-3s for transmission at a trunk transmission rate of typically 560 Mb/s to 4.8 Gb/s per fiber pair. Any overhead signaling, orderwire, alarm, or control data between POPs or re-generator sites is combined with the channel signals at the POP terminals. The composite digital data stream is encoded for transmission, with error encoding inserted, and converted to a lightwave signal in the optical transmitter. The signal is transmitted on a single fiber pair for each high-level multiplexer.

The protection switch is a 1:N switch that protects against terminal and repeater outage along the terminal-to-terminal span or path. The degree of redundancy is designed to achieve the necessary equipment availability for each span and allow online repair without span outage. The protection switch does not protect against cable cuts, which usually sever all fibers at once.

At a regenerator, the optical signal is detected and regenerated, that is, the optical on-off signal is detected and converted to an electrical signal and that signal is retimed and restored as a square-wave binary signal. This signal is then input to an optical transmitter and converted back to an optical signal. Overhead information, including error and alarm monitoring, is dropped and inserted as well.

With asynchronous equipment, signals are combined and extracted (multiplexed and de-multiplexed) all at once at standard DS-1 or DS-3 rates (1.544 Mb/s or 44.736 Mb/s). If an individual DS-3 channel is to be added or extracted at any repeater or end point along the trunk all DS-3 channels must be demultiplexed and remultiplexed. With synchronous SONET equipment a single DS-3 channel (imbedded in the SONET 51.840 Mb/s signal) can be dropped or added without demultiplexing all the other signal traveling along the fiber. This difference will be discussed in a later section.

An alarm, control, and orderwire system supports network operation. Network management signals are generally carried in part as terminal-to-terminal overhead (section overhead) multiplexed in with the information channels, and in part as alarm-and-alert information carried on separate overhead channels (line overhead) multiplexed into the high-level data stream.

Alarms and system status are generally sensed at all levels within the equipment. Some of the typical fiber terminal alarms and status indicators include loss of signal, loss of frame synchronization, BER above threshold, "blue" signal (generated when no signal is present), individual module or channel failure, power supply failure, backup power and protection status, and terminal configuration. Housekeeping alarms can also be inserted such as door open, primary power failure, motor generator activated, temperature sensors, and so on.

An orderwire channel is generally included for communication and control signals. Control signals may include communications between protection switching equipment, repeater hut security controls, control of backup generators, and so on. Voice orderwire is also provided from site to site along a selection for troubleshooting.

7.3.2 Asynchronous Transmission Standards

At the electrical interface, asynchronous fiber terminals have generally been standardized at the DS-3 signal interface of 44.736 Mb/s, at least in the United States. Lower speed transmission terminals (50 Mb/s to 150 Mb/s) will often have a DS-1 interface standard (1.544 Mb/s). Both standards are well defined by BELLCORE.

At the optical interface, the input to the fiber, no standards exist for asynchronous multiplexing fiber terminals. Each manufacturer uses its own signaling format, degree of overhead, error encoding, and synchronization; therefore, the output data rates all differ. Terminal equipment must be of the same manufacturer and type end-to-end in any optical span.

Some pseudostandardization exists as to the level of multiplexing (number of DS-3s per fiber) incorporated on commercial fiber terminal products. The various transmission rates associated with those levels are illustrated in Fig. 7.8 as the hori-

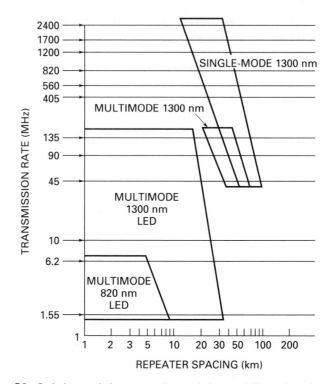

Figure 7.8 Optical transmission rates and transmission capability typical of available asynchronous product.

zontal lines (vertical axis). Those rates at 45 MB/s and above are formed by combining multiple DS-3 (44.736 MB/s) channels.

Fig. 7.8 also illustrates the performance capability of fiber-optic digital trunking equipment based on today's typical product. Actual performance will vary depending on manufacturer and outside plant components used.

7.3.3 Synchronous Transmission Standards

The Synchronous Fiber Optic Transmission standard [2] was developed by the American National Standards Institute (ANSI) and the Exchange Carriers Standards Association (ESCA) as a means for standardizing both the electrical and optical interfaces for future transmission systems. It provides for a standard optical interface so that equipment of different manufacturer's can interface optically, fiber to fiber. It also provides for a standardized electrical multiplexing approach that can combine and carry multiple signal formats that are today incompatible. It multiplexes and serves the North American hierarchy of signals used by digital telephone systems today (DS-0, DS-1, DS-1C, DS-2, DS-3) as well as Integrated Services Digital Network (ISDN) and the fourth hierarchical level DS4NA, which is an international standard.

The standard also defines a set of overhead or network management channels that will potentially permit user network management information to flow over public SONET networks [2].

Fig. 7.9 illustrates the structure of a SONET system consisting of the path layer, line layer, section layer, and photonics layer. Some equipment does not require all four layers. For example, in a repeater only the photonics and section layers are present. Also, terminals that do not drop and insert signals do not require the path layer.

7.3.3.1 Path Layer

The path layer is the highest-level layer. It provides the signal grooming and multiplexing necessary to combine a signals to be transmitted (called "services") together with any path overhead (POH) signals into a single digital bit stream called a synchronous payload envelope (SPE).

A path is defined [2] as a logical interconnection between a point at which a standard SPE frame is assembled and the point at which it is disassembled. The path layer communicates within itself end-to-end via the path overhead (POH).

Signals with a data rate much lower than the lowest line layer STS-1 rate (51.480 Mb/s) are carried in virtual tributaries. These are combined with signals from other virtual tributaries in the path termination equipment to make up the signal called the payload envelope (SPE) at the STS data rate and format compatible with the line layer equipment. The STS-N multiplexing rates and virtual tributary rates are described in Table 7.1.

Although most of the multiplexing (or mapping) converts sub-STS-1 signals into STS-1 payload envelopes, a superframe STS-Nc format is also specified to carry such multiple STS-1 services as a broadband ISDN H4 channel.

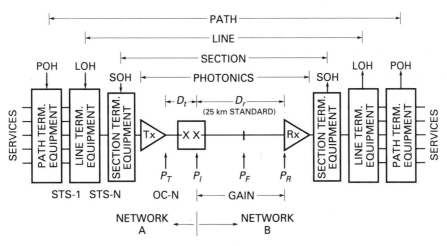

Figure 7.9 SONET model. (From Ref. 1; copyright 1990, Prentice Hall Inc.)

TABLE 7.1 SYNCHRONOUS MULTIPLEXING HIERARCHY

STS level	OC level	Line rate (Mb/s)	Standard payload mapping
STS-1	OC-1	51.840	DS3 at 44.736 Mb/s
			SYNTRAN at 44.736 Mb/s
STS-3	OC-3	155.520	DS4NA at 139.264 Mb/s
STS-9	OC-9	466.560	
STS-12	OC-12	622.080	
STS-18	OC-18	933.120	
STS-24	OC-24	1244.160	
STS-36	OC-36	1866.240	
STS-48	OC-48	2488.320	
Virtual tributary rates			
VT1.5		1.728	DS1 at 1.544 Mb/s
VT2		2.304	2.048 Mb/s
VT3		3.456	DSIC at 3.152 Mb/s
VT6		6.912	DS2 at 6.312 Mb/s

Source: ANSI T1.105, March 9, 1988.

7.3.3.2 Line Layer

The line layer multiplexes path layer SPEs and line overhead into an STS-N level bit stream for transmission over the Section and Photonics layers. LOH is accessed at all Line Termination points where STS-N signals are created.

The lowest-level STS is the STS-1 at a rate of 51.840 Mb/s. The STS-1 frame consists of 810 bytes (6480 bits) with a frame length of 125 μs (8000 frames per second). Twenty-seven of the bytes are overhead (LOH) and 783 are for the payload envelope (which is mostly data with 9 bytes being for POH).

The line layer overhead LOH communications channel bandwidth is 192 kb/s and it uses a packet protocol.

In addition to a multiplexing function the line layer provides Line maintenance and protection. 1:1 and 1:*N* protection switching are specified for this layer. Specified values are from 1 to 14 channels that can be switched to an optical protection channel.

7.3.3.3 Section Layer

The section layer transports STS-N signals across the physical medium, by mapping STS-N signals and section overhead (SOH) into a bit stream for the photonics layer. The section layer provides framing and scrambling of the signal in a manner compatible with optical transmission by the photonics. It also performs error monitoring.

The SOH data communications channel bandwidth is 576 kb/s. It uses a packet protocol. SOH is used for communicating operations, provisioning, and administration information as well as local orderwire.

7.3.3.4 Photonics Layer

The photonics layer is the fiber optics transmission layer. It converts the STS-N electrical signals into synchronous optical carrier signals (OC-N) at the same bit rate as the electrical STS-N signal.

The specifications for power levels, system gain, wavelength, physical interconnection, and optical pulse shape are provided in T1X1/87-128R1 [2]. Some of the characteristics of the photonics layer as recommended by the standard include:

Fiber: single mode

Operating window: 1310 nm for SMF; 1550 nm for DS-SMF

Laser center wavelength

    1310 nm MLM laser = 1270 nm − 1340 nm distance dependent

    1310 nm SLM laser   = 1280 nm − 1340 nm for sections <40 km

    1550 nm MLM laser = 1525 nm − 1575 nm distance dependent

Spectral width

    1310 nm MLM laser = 3.5- to 30-nm rate and distance dependent

    1310 nm SLM laser   = < 1.0 nm

    1550 nm MLM laser = 3.5- to 30-nm rate and distance dependent

Section performance

    10^{-9} BER for 40 km or less

    Standard section: 25 km

Connectors

    Optical return loss: > 20 dB

    Standards: EIA RS-455-XX and EIA 4750000-A

These parameters are provided as general guidelines only. The reader should refer to the standard before proceeding with a design where more specific values are given as referenced to specific line rates and transmission distances [2]. This is also necessary because standards change from time to time, particularly when they are in draft form.

7.4 MANs

A MAN is one that transports signals within a city, between buildings, distribution nodes, or provides local distribution for WAN traffic.

From an applications standpoint, MANs generally involve

1. Subscriber loop communications
2. Access from a customer premise to an interexchange carriers POP
3. Private networks, interconnecting office buildings, business complexes, or LANs
4. CATV networks involved in the transport of entertainment and educational traffic

7.4.1 Application of Fiber in MAN

Fiber optics has gained popularity as a transmission medium in the MAN environment primarily because it is low in operating cost (when its capacity is used) and very simple for commercial businesses to implement without a heavy technological or resource investment.

Fiber is a high-performance transmission medium relative to other MAN transmission approaches. Once the fiber is selected and installed, concern for the transmission medium is minimal since there exists no outside plant electronics and the characteristics of the cable are stable with time, temperature, and electromagnetic interference. Microwave systems are limited by noise interference, environmental fade, physical blockage, and FCC regulations. Coax systems suffer from interference ingress, changes in characteristics with temperature, corrosion, and degradation of plant electronics. Twisted pair is constantly susceptible to changing amounts of interference and cross talk.

Fiber systems are also very simple, generally consisting of fully modular end terminals with no repeaters. Installation is simplified by the small, light, rugged nature of the cable, making fiber easier to install than other cable types. The cable can be overlashed in 2- to 3-km spans on existing aerial plant, plowed directly into the ground in 5- to 10-km spans, or pulled continuously (using special techniques) through 2 to 3 km of duct.

Maintenance and operation is simplified because the equipment is generally modular, requires no setup or periodic adjustment, and generally contains microprocessor based self-diagnostics. One to 3 years mean time between maintenance action and well over 5 years mean time between unprotected outage are common. Other than for repair of physical damage to the cable, plant maintenance is nonexistent.

Fiber finds application in the MAN when transmission capacities are large, when transmission distances are longer than achievable with twisted pair, and when environmental conditions (EMI, temperature, humidity) dictate. Fiber is generally applied to the transmission of:

1. High-speed data or data packets (at the 1- to 10-Mb/s rate or above)
2. PCM voice traffic where multiple DS-1 circuits are required
3. High-quality broadband video transport for broadcasters or CATV supertrunks
4. Systems requiring a high degree of availability and path redundancy between a few locations.

Fiber is less applicable where

1. Data rates are low (below 1 Mb/s)
2. Distance is within the range where twisted pair can better perform the function
3. Single voice or data circuits are involved
4. The application requires distribution of low capacity signals to a large number of users (such as CATV distribution, telephone, etc.)

7.4.2 Network Architecture

For the most part, MAN transmission will be aimed at the transport of voice, data, and video information, within dedicated point-to-point circuit bandwidth, between buildings, or between clusters of buildings. The public MAN network will contain a channel cross-connect node and network control center, the central hub, where network monitoring and provisioning occur.

Fig. 7.10 illustrates four basic topologies for communication between locations in a cluster: star, ring, bus, and tree.

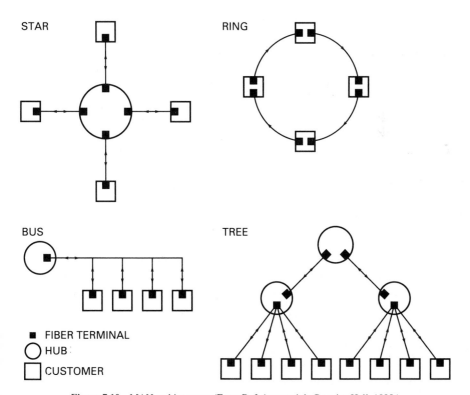

Figure 7.10   MAN architectures. (From Ref. 1; copyright Prentice Hall, 1990.)

7.4.2.1 Star

Star architecture provides direct point-to-point service to customers on dedicated lines emanating from the central hub. Multiple customers may be physically attached to the cable, forming the "spokes", but the channels are dedicated point-to-point by virtue of separate fiber pairs and terminal pairs dedicated to a particular customer. Star architecture is common to the copper pair Telco Central Office Local Loop in today's environment. In the event of a cable cut all customers on that cable are out of service unless a backup cable along a different route is provided.

One means of avoiding the vulnerability of a cable cut is to make the fiber run for each arm of the star diverse, that is, two fiber paths between the central hub and each customer. This may sound expensive but if properly planned the cable can be routed between all customers to form a ringlike path, while the fibers are spliced in a manner that creates two point-to-point diverse paths, one in each direction around the ring. This will be described later in more detail.

The advantage of the star is that fiber and terminals are dedicated to customers and therefore can be customized to the requirements of that customer. Star architecture is also ideal for passive optical hubs or splitters. The disadvantage is that a pair of fibers and a pair of terminals are required for each customer. Diverse routing requires that the pair exist in both paths and that the terminal equipment be redundant as well.

7.4.2.2 Ring

Ring architecture provides a continuous loop of fiber interconnecting all nodes (customer locations and central hub) in the network. The bandwidth of the fiber ring is shared between all customers, who drop and insert their channels at the nodes. The ring is generally configured with the flow of information around the ring in one direction. If the cable is cut or a terminal fails, however, all communications on the ring would cease. To protect against this the electronics is made redundant and a counter-rotating ring is configured in the same cable on another fiber. If a cable or terminal failure occurs, the ring reconfigures and loops back on itself, sending information around the other way so that information flows between connected operating nodes. The operation of such a ring will be further discussed in a later section. Circuits to any one customer are logically connected in a point-to-point fashion but electronically connected as a TDM loop.

The advantage of the ring architecture is that it uses the least amount of electronics and fiber pairs. Multiple customers can be connected with a single pair of fibers. The disadvantage is that all customer's traffic flows through the electronics at each node. This can have security implications. It also means that traffic collected around the ring is additive at each node, thus extending capacity requirements of any one node even though that node may drop or add very little traffic. The disadvantage of the cost of high-speed fiber terminals may outweigh the slightly higher quantity of terminal equipment of the point-to-point star approach.

7.4.2.3 Bus

The bus shares the bandwidth of a common cable by using multiple access approaches that channelize the cable in time or frequency, and drop/insert channels at customer terminal points. It operates much like the ring except that it is not physically closed. The most common physical bus architecture in a MAN is that used with CATV institutional networks where subscribers are assigned different frequency channels on a single coaxial cable. Channels are cross-connected at the master node (head end) by frequency translation from a transmit frequency band to a receive band.

The main advantage of the bus architecture is flexibility and ease of connection. A major disadvantage of this architecture is that if a cable is damaged, all subscribers beyond that point are affected. With fiber the greatest disadvantage is that optical coupler tap loss is too great and coupler cost too high to make this practical.

7.4.2.4 Tree

In the tree architecture, customers are connected in a hierarchical fashion to local nodes that can perform circuit concentration and switching functions. The concept of the local node permits high quality supertrunking of channels between the central node and the local nodes; it also permits a low capacity but more economical distribution from the local nodes to customers. This architecture is common to the public switched telephone network where local serving office switches are connected in a hierarchical fashion to larger switches. This architecture is also common to the CATV industry for TV broadcast.

The advantage of tree architecture is the same as that of the star with the added provision that capacity and cross-connect requirements can be distributed for better economies. The disadvantage is that a local node outage or cable outage affects all customers and nodes downstream.

7.4.3 STAR Distribution With Fiber

Figure 7.11 illustrates the configuration of a single cable emanating from a master hub point. In the star configuration each transmission channel is carried on a separate pair of fibers, one for transmit, one for receive. Cross connection of channels from one node to another involves electro-optical conversion at the central hub. Termination to the fiber cable involves splicing into a pair of fibers within the cable at an appropriate splice point. This must be carefully planned since all fibers must be spliced at each splice point unless special splicing practices are employed. For this reason multiple terminals (and fiber pairs) are often dropped from each splice point. Terminals can be added as long as the fiber count permits.

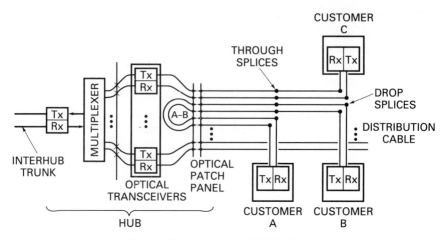

Figure 7.11   Nonredundant STAR distribution.

7.4.4 Path Redundant STAR Distribution With Fiber

Figure 7.12 illustrates a convenient means of providing total equipment and path redundancy with star architecture. The interconnects are essentially point-to-point as
with the star, but duplex terminal equipment and diversely routed fibers bundled
within the same cable provide path redundancy. In this configuration all signals are
brought to the master node and cross-connected there. Since the terminals do not

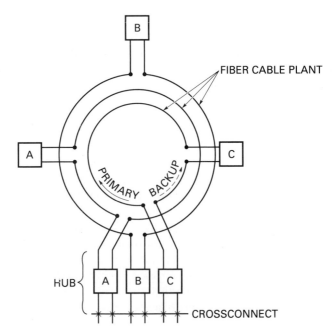

Figure 7.12   Path redundant STAR architecture.

share fiber bandwidth as with the ring, additional terminal capacity can be added to those sections that have higher traffic loads, without affecting the capacity limits of other terminals. The disadvantage is that this configuration uses two fibers in the entire cable (a pair in each direction) for each customer or node on the network.

A functional block diagram of the terminal electronics that performs the star configuration redundant trunking function is illustrated in Fig. 7.13a. The terminal generally contains channel multiplexers to combine a number of low-speed channels into a high-speed channel for transmission on the fiber pair. Additional multiplexing at lower rates can be added externally to modularly customize the terminal to the application. The multiplexer equipment, fiber transmitters and receivers and fiber pairs are redundant and switched by an automatic protection switch in the event of degradation or failure of the operating channel. Generally the fiber terminals contain 1:N low-speed (input channel) protection and 1:1 high-speed (optical transmitter receiver) protection.

For MAN applications, fiber terminal equipment can be purchased with built-in multiplexers that combine from 4 to 28 DS-1s or 1, 3, or 12 DS-3s. For premise equipment, typically a 150 Mb/s unit is used that has the ability of growing modularly from a capacity of one DS-3 to three DS-3s. Figure 7.13b represents a unit that can expand from 1 to 3 DS3s.

7.4.5 Active Fiber Optic Ring

The active ring, illustrated in Fig. 7.14, is an ideal structure from the standpoint of redundancy and fiber economy. It requires only one fiber around the ring to transmit to all nodes, plus a second for counter-rotating redundancy. The active (repeating) ring capacity suffers, however, in that each terminal in the ring must handle the additive capacity of all channels inserted at each node around the ring. For example, signals passed between locations A and D must also be passed by terminals B and C. This situation can quickly reduce all capacity of the ring requiring higher speeds at each node.

Another practical issue that results when using the active ring in a public network, is the placement of terminal nodes within the private premises of each user. These nodes carry (and loop through) traffic from other users. The ring is therefore only reserved for private networking or as an off-premise trunk in a public network.

In the ring structure in Fig. 7.14 consist of multiplexers (MUX) and demultiplexers (DUX), which drop and add channels. Through channels are simply cross connected at the input/output of the multiplexer/demultiplexer pair (unless SONET is used whereby only the local channels are affected). The MUX/DUX pair is then served by a pair of primary optical transmitters and receivers that, in the figure, route all information flow clockwise around the ring. Dedicated node-to-node channels are dropped and inserted as they pass through their assigned terminating nodes.

In the event of a cable or node failure, transmitter/receiver pairs are located at each node with an associated bypass switching function. The function is to loop the signal around at the node just downstream of the failure point and send it in a counter-rotating direction around the ring, bypassing all other nodes, until it reaches the node just upstream of the failure point. The switching function is shown conceptually here

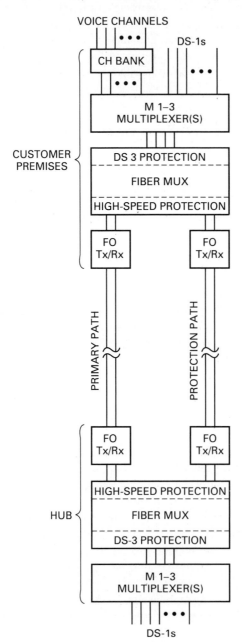

Figure 7.13(a) Path redundant fiber optic distribution equipment[1].

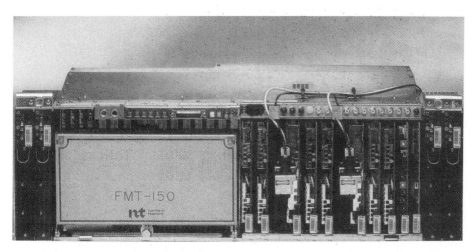

Figure 7.13(b)   Fiber Optic Customer Premise distribution equipment with 3 DS-3 capability (courtesy Northern Telecom).

as a double-throw switch. In reality switching is performed electronically [3]. Each product will implement it a little differently.

Loopback control is the function of the terminal controllers and imbedded ring control logic, which maintains status and configuration communication between all nodes. In the event of a fault condition each node analyzes the condition (loss of signal and direction), takes appropriate local switching action, and notifies the systems control function of the situation through the imbedded overhead communications channel. Whether central or distributed in each node's controller, the system control reconfigures all nodes as appropriate to the fault condition and location.

Figure 7.14 illustrates a failure condition with a cable cut between node B and node C. If we follow channel A to C (AC) through the network it will illustrate the protection arrangement. Channel AC enters the MUX at node A where it is transmitted on the primary fiber ring to node B where it is looped through. Node B has sensed the upstream failure (absence of signal of fault condition from C) and has switched all outbound traffic to the protection transmitter. Channel AC therefore is routed counterclockwise back to node A. Since there are no failures adjacent to node A its protection path remains in the normal bypass mode, acting like a repeater, and passes channel AC through to node D. Node D treats the channel in the same manner and passes it on to the node C protection receiver. Node C has recognized the downstream cable cut (no signal from node B) and has activated its downstream bypass switch which routes signals from the protection receiver into the demultiplexer. The DUX at node C recovers the channel AC signal.

Figure 7.15 illustrates ring equipment which operates at 565 Mb/s.

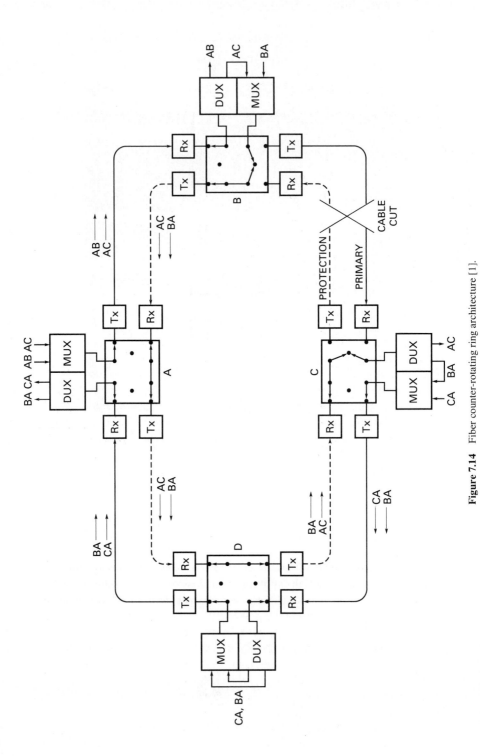

Figure 7.14 Fiber counter-rotating ring architecture [1].

Figure 7.15   Fiber optic counter-rotating ring equipment (courtesy Northern Telecom [3]).

7.5 CABLE TV APPLICATION FOR FIBER

7.5.1 Video Applications in the MAN

Although data and voice transmission has sustained the most interest for fiber transmission in the MAN, the requirement for delivery of high quality TV to subscribers
in large metropolitan areas has stimulated CATV application. High quality video
transmission on fiber has a definite application in areas where conventional methods
(coax and microwave) are not totally satisfactory in performance or cost. Video applications most often include 1) trunking of broadcast signals to cities across the na-
tion and 2) distribution of CATV channels to subscribers.

Short-haul broadcast trunks generally require only a few (1, 4, etc.) channels to
be transmitted with NTSC* weighted signal-to-noise ratios of 62 to 67 dB over distances of 5 to 10 miles, to reach a carrier, satellite, microwave site, or second studio.
Historically these systems have been coaxial or microwave, or if fiber, single channel
per fiber using FM or digital PCM modulation.

*NTSC - National Television System Committee

Long-haul trunking is generally done by satellite where the full 6-MHz bandwidth per channel is carried on an analog or FM modulated carrier. Fiber can easily carry the video bandwidth, but requires the signal to be digitized, which is an expensive process and requires approximately 90 to 150 Mb/s per channel. The signal is often compressed to fit within a standard DS-3 format (44.736 Mb/s) for trunking over the type of digital telecom lines described in section 7.3. This of course requires video compression that adds greatly to the expense of the video encoder.

CATV trunking requires a weighted signal-to-rms-noise ratio of about 52 to 55 dB peak-to-peak, white-to-blanking (ppwb/rms)w. CATV trunks in a metropolitan area generally carry 50 to 100 of channels over distances of 10 to 20 miles between the head end and remote distribution nodes. These signals are then distributed from each node to thousands of homes, arriving with minimum signal-to-noise requirements of between 40 to 45 dB (ppwb/rms)w. Although the signal-to-noise requirements for CATV are not as stringent as for broadcast the combination of SNR, channel capacity, and low cost makes the CATV requirement the most challenging for any transmission technology.

7.5.2 CATV Application of Fiber

A typical CATV network is illustrated in Fig. 7.16. The potential application areas for fiber are 1) the feed trunks between signal source and head end, 2) the supertrunk between the head end and the hub and 3) the distribution network. Unless clever architectural design is used, fiber is limited in its ability to compete with coax in the distribution portion of the system. In the conventional tree architecture shown, the fiber, connectors, and splitting components are too costly and power losses too great in comparison with coax with today's technology. Section 7.9 and Fig. 7.27 illustrate the optical loss limitations of the tree type distribution network with passive optical taps.

Historically coax and microwave have been the only technologies that have been accepted by the CATV industry. Up until single mode fiber technology became practical, fiber was too expensive for CATV. When single mode fiber became available at production volumes and low cost in 1983/1984, the picture changed. Development work and field trials [4,5] illustrated that multichannel singlemode fiber supertrunks were lower cost and higher quality than FM coaxial supertrunks used in the larger CATV networks. Some successful installations followed by two of the largest multiple systems operators, Warner Amex [6] and ATC [7]. Perceiving a new market in CATV supertrunks, video fiber optic product development followed with a number of manufacturers. Fiber began to be the medium of choice for supertrunking, on a cost/performance basis, where distance was great and signal quality had to be high [8,9].

In the system referenced above, FM modulation was used as illustrated in Fig. 7.17. It had the disadvantage that a complete set of vestigial sideband (VSB) modulators, or CATV headend, was needed at each receiving hub to place the video signals into their proper channel frequencies on the coaxial distribution system. A method for

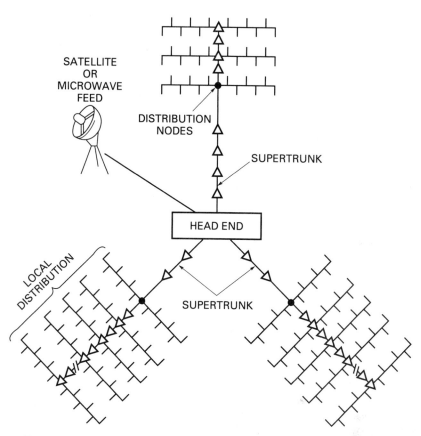

Figure 7.16   Large metropolitan-area CATV network using supertrunking and distribution nodes [1].

direct VSB transmission over fiber was therefore needed. In 1984, this author at Warner Amex demonstrated the ability to transmit from 4 to 6 channels of direct analog (VSB) video over 10 km or more of singlemode fiber with 50 dB or greater weighted signal-to-noise ratio. Very linear lasers from General Optronics and Laser-tron were required to reduce the intermodulation products. This was significant in that it permitted the video channels to be transmitted over fiber in the same format and channel frequencies as they appear on the coaxial cable and into the home. The method in Fig. 7.19 was used, and as this figure shows, it eliminates the need for additional FM or PCM signal processing as well as the need for an additional VSB headend at the receiving points.

In the years hence, production quality lasers have become more linear and singlemode lasers more available. Using singlemode distributed feedback (DFB) lasers, for example, Ipitek (a division of TACAN Corp) markets an analog system that transmits up to 60 channels per fiber, with a 9 dB loss budget (20 km or greater) [10].

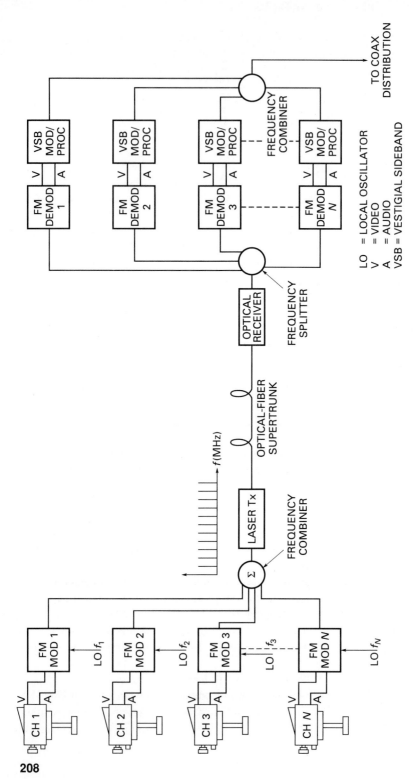

Figure 7.17 FM/FDM fiber-optic video trunk for CATV [1].

LO = LOCAL OSCILLATOR
V = VIDEO
A = AUDIO
VSB = VESTIGIAL SIDEBAND

The transmitter for a dual 600-MHz trunking system is illustrated in Fig. 7.20. A carrier-to-noise ratio of over 50 dB is advertised at 60 channels and a ratio of 55 dB at 20 channels per fiber. Colony Cable indicates that with proper laser selection, excellent results with 80 analog channels per fiber are possible. The perfection of direct analog transmission on fiber permits the economical introduction of fiber into parts of the distribution plant.

7.5.3 Fiber Architecture for CATV

Using fiber in CATV networks saves greatly on cost, amplifier equipment, maintenance, pole space, and power requirements. Depending on the modulation approach used and performance considerations, a single fiber cable with from 2 to 6 fibers can carry 100 channels of video over 10 miles. The equivalent AM or FM coaxial implementation for 100 channels requires (in one case, reference 6) four 3/4-inch cables and 56 amplifiers for a 10 mile span. In comparative performance tests [6], FM coax delivered only 53 dB weighted SNR and the AM coax less than 50 dB weighted SNR, while the fiber delivered over a 60 dB weighted SNR.

In the longer supertrunks where many very high quality channels are required, the video signal usually requires further processing in order to reduce the required level of the transmitted carrier while overcoming the effects of noise and intermodulation distortion. Whereas the received video signal must maintain a peak level 60 to 70 dB higher than the rms noise, the composite carrier and signal transmitted over the fiber must be no greater than 20 to 30 dB greater than the rms noise due to the fact that there is not enough optical power margin available. Signal processing (or modulation and encoding) achieves this reduction of the transmitted carrier and reconstructs the higher SNR signal at the receive end. This is of course done at the expense of wider bandwidth, but fiber systems generally have more bandwidth than they do signal power, so this is an appropriate trade.

The most popular signal processing approaches are 1) frequency modulation of the video signal in each channel, then frequency division multiplexing of multiple channels (FM/FDM), and 2) digitizing each video channel using pulse code modulation open time division multiplexing the channels onto the fiber (PCM/FDM). These methods are illustrated in Figs. 7.17 and 7.18.

For the same signal performance, uncompressed linear PCM can outperform AM and FDM in distance and its ability to be repeated without signal degradation. A video transmission product from Ipitek (R) is advertised as able to carry up to 10 uncompressed channels per fiber with 8-bit encoding at 1.25 gigabits per second over a 20 dB loss budget [10]. Direct analog systems will generally have less than a 10 dB margin for similar or less signal performance. Twelve to 16 channel FM systems may also indicate 10 to 20 dB margins, but will lose at 3 dB performance if repeated. Although PCM/TDM outperforms analog the relative cost advantage goes to analog. Video compression widens this disadvantage even more. This should change over time as volume production drives down the cost of digital encoding.

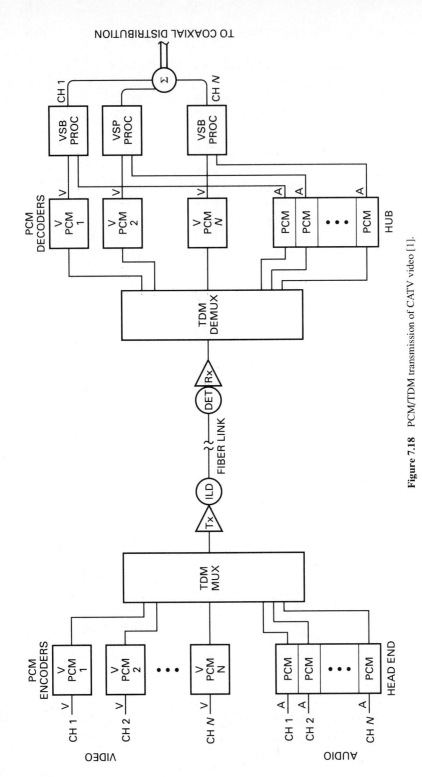

Figure 7.18 PCM/TDM transmission of CATV video [1].

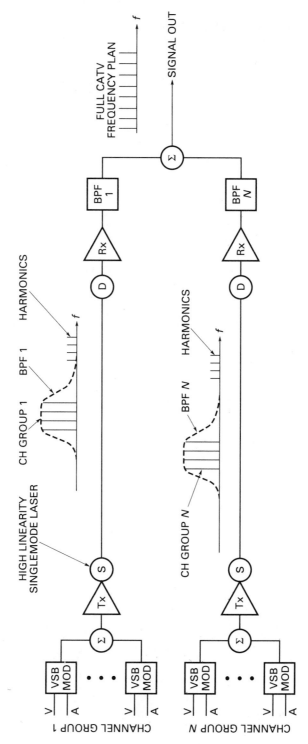

Figure 7.19 Vestigial sideband AM fiber-optic transmission of CATV video.

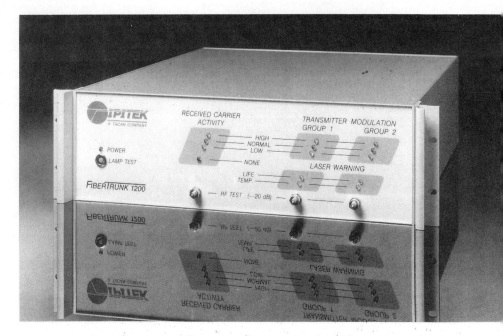

Figure 7.20   FiberTrunk™ 1200 High Performance Broadband Fiber Optic AM Transmission System (courtesy IPITEK).

As laser linearity and power improves, the use of direct analog modulation may totally supplant FM or PCM in all but the most high quality long-distance super-trunks. Cost is about the same or less than AM coaxial distribution trunks, particularly when maintenance and power is considered and performance is better.

By using the proper architecture, fiber can enter the distribution plant today, and perhaps even the home. Figure 7.21 illustrates an architectural approach that brings fiber into the distribution plant up to the point of the drop to the home. At this point coax is generally used due to the lower cost than a drop fiber system. LED-based fiber drops could be used if the application outweighed cost considerations. The products sold by Ipitek [10], for example, permit an all-analog system to be constructed in this manner, from headend to drop node.

In the mid 1980's fiber to the home was actually installed by this author and others, using Times Fiber mini-hubs in a similar architecture whereby the drop node was an active converter. The signals were transmitted to the drop mini-hub over fiber or coax. The mini-hub consisted of the CATV converters that, instead of being with the TV, were within the outside hub. The fiber drop carried the channel of choice into the home, and the channel selection command signal from the home to the mini-hub converter.

In the future, coherent optics and integrated optics may make direct analog approaches even more popular since this technology will eventually permit 100 channels to be wavelength division multiplexed on a single fiber. Today the best produc-

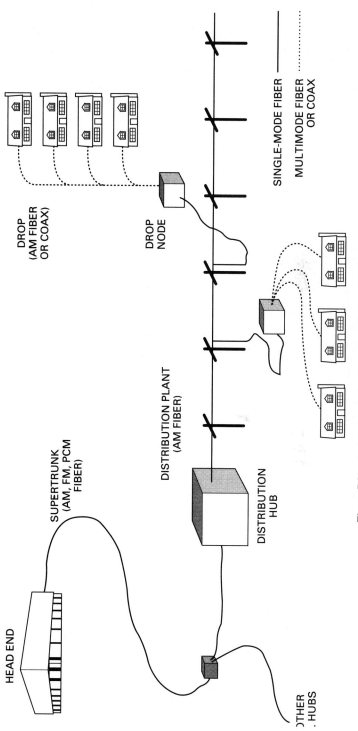

Figure 7.21 Architecture for implementing fiber in a CATV network.

SINGLE-MODE FIBER ——————

MULTIMODE FIBER ··············
OR COAX

DROP
(AM FIBER
OR COAX)

DROP
NODE

DISTRIBUTION PLANT
(AM FIBER)

DISTRIBUTION
HUB

SUPERTRUNK
(AM, FM, PCM
FIBER)

HEAD END

OTHER
HUBS

tion communications lasers emit light with a broad band of frequency components. With tomorrow's coherent optics technology, each laser carrier frequency will be virtually pure (near single frequency) allowing it to be treated like a radio wave (direct FM modulation of a laser optical waveform for example). Tuning a CATV channel will become a matter of tuning a variable frequency laser at the receiver to the optical frequency of the desired channel. This is the same approach used today with the CATV converter on your TV today, except that the tuner operates at radio frequencies instead of optical frequencies.

7.6 LOCAL AREA NETWORKS

A LAN is a multiterminal, multiple access communications systems operating within a somewhat closed or private environment and within a limited distance and fixed transmission data rate. A LAN is generally a private computer, workstation, and server-based data transmission network operating within the office environment between hosts and workstations or between PCs. The LAN has historically spanned a single work group in an office, a single floor, or a single building. With recent bridge, router and repeater technology the LAN is now spanning the metropolitan and wide area networks to interconnect with other LANs in a networking environment. New terminology has been attributed to these LAN networks. LANs which span buildings in a complex may be known as a Campus Area Network (CAN). The wide area interconnection between local LANs or CANs is called a LAN Interconnect Network (LIN).

7.6.1 LAN Characteristics

A LAN is characterized by the fact that, unlike trunking systems, the transmission medium and the bandwidth are shared among the stations, through some multiple access means. Multiple access mechanisms are both physical and logical. Optical couplers, for example, represent a physical multiple access device and multiple access signal protocols represent the logical rules for sequencing the communications between stations.

The two multiple access protocol methods most often used with LANs are as follows (see Fig. 7.22):

a) Carrier Sense Multiple Access /Collision Detection (CSMA/CD) described in IEEE standard 802.3 [12];

b) Token Passing described in IEEE standard 802.5 [13, 14] and ANSI FDDI standard [15, 16].

With CSMA/CD (more commonly known as Ethernet), each station listens for anyone else's transmissions (senses carrier) before it attempts to transmit on the LAN and will wait to transmit until it is clear to do so. If two stations accidentally transmit at the same time, then a collision will occur, as illustrated in Fig. 7.22a. The result is

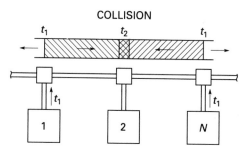

t_1 = TRANSMISSION BEGINS AT TERMINALS 1 AND N SIMULTANEOUSLY
t_2 = COLLISION OF LEADING EDGE TAKES PLACE
t_3 = COLLISION DETECTED
t_4 = TRANSMISSION CEASES AT ALL TERMINALS
t_5 = TERMINALS WAIT AN INDETERMINATE AMOUNT OF TIME TO TRANSMIT
       AFTER SENSING THAT THE BUS IS CLEAR

(a)

t_1 = TERMINAL N GIVES UP TOKEN AND PASSES IT TO TERMINAL 1
t_2 = TERMINAL 1 TAKES TOKEN AND TRANSMITS INFORMATION INTENDED
       FOR TERMINAL 3
t_3 = TERMINAL 3 ACCEPTS THE INFORMATION AND PASSES ON THE TOTAL
       DATA STREAM
t_4 = TERMINAL 1 SENSES ITS OWN TRANSMITTED DATA RETURNING, DELETES
       THE DATA, AND FREES THE TOKEN, PASSING IT ON TO 2

(b)

Figure 7.22 Contention-based multiple-access LAN approaches. (a) CSMA/CD.
(b) Token passing.

a distorted "collision" signal that is sensed by all stations. All stations will then cease transmission for slightly different time periods and then try again.

With token passing (better known as token ring) each station is sequentially given an opportunity to transmit by being offered a little electronic packet of data bits known as a token. As illustrated in Fig. 7.22b, the token, which gives a station the right to transmit on the LAN, is passed around the network from station to station. If a station receiving the token has nothing to transmit, then it will regenerate the token and pass it on to the next. If a station has transmitted, it will regenerate the token and pass it on to the next station once it sees its own signal coming back around. If for example station 1 in Fig. 7.22b receives the token and wishes to transmit to station 3 it will transmit, whereby station 3 will receive and repeat the message back to station 1. Station 1 will then release the token to the next station 2.

The role of fiber in the LAN environment is to replace the more conventional mediums of coax and twisted pair, to reduce electromagnetic interference, and to increase data rate and distance. Most LAN standards have concentrated on coax and twisted pair for transmission rates below 16 Mb/s. The major impact that fiber has had on LANs is to provide the capability for high-speed LANs running at 10 Mb/s, 16 Mb/s, and 100 Mb/s or above. The FDDI standard was developed around the unique capability of fiber. It provides for a 100 Mb/s token ring LAN that can be extended to 100 km.

Figure 7.23 illustrates hypothetically how these various LANs might be used

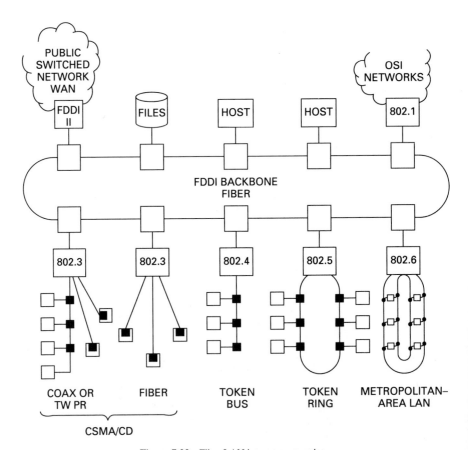

Figure 7.23   Fiber LAN interconnect options.

and interconnected to serve various applications within a data network. The 100 Mb/s FDDI LAN is a natural to form a network backbone or a backbone within a building or office complex. It is compatible with all other 802 LANs and can therefore gateway into them. Extension of the LAN across a WAN is possible using the FDDI-II Hybrid Ring Control (HRC) [17], SONET, or frame and cell relay technologies.

The 802.3 through 802.5 LAN standards are structured more for the office size LAN. They are ideal for interconnects between PCs or between host processors and peripherals. The 802.5 token ring standard twisted pair, and the 802.3 CSMA/CD specifies standard coax and twisted pair [12, 18]. Both standards are being updated for fiber [19, 20]. Data rates of up to 16 Mb/s have been established for twisted-pair LANs; however, it is at this point that the use of twisted pair becomes troublesome and fiber may be the better choice for LAN cabling.

Fiber may also have an impact on the IEEE 802.6 standard for Metropolitan Area Network LAN structures, particularly as it might be used in fiber to the home applications [21].

7.6.2 FDDI Fiber Optics LAN

Although some standards activity is underway for adding fiber to IEEE 802.3 [14], perhaps the most significant token-passing LAN standards activity involving fiber has been the ANSI X3T9.5 Fiber Distributed Data Interface (FDDI) [15]. This standard recommends a 100 Mb/s, dual counter-rotating token-passing ring. FDDI can support a sustained data transfer rate of about 80 Mb/s. The standard is designed to support 1000 physical connections (500 terminals) and a total fiber path length of 200 km (100 km dual ring). The information supplied in this section is derived from the July, 1988 draft standard [15, 22].

Figure 7.24 illustrates the interconnection of stations on a dual counter-rotating FDDI ring. The stations are serially connected in a ring forming a closed loop, each station connecting to both a primary ring and a secondary ring. Information is transferred from active station to active station in one direction around each physical ring. The primary ring consists of an output, primary out (PO), and input, primary in (PI), and the secondary ring an output, secondary out (SO), and input, secondary in (SI). The internal electrical connection of each station is controlled by insertion and removal commands from a LAN control system known as the SMT layer. An optional optical bypass switching arrangement is also specified for FDDI such that a failed or inactive station can be bypassed optically under the control of the SMT.

The SMT layer supporting all systems management applications, accumulates operating statistics, and initializes each station. In the event of an outage or a station going inactive the SMT reconfigures the ring by 1) commanding the station bypass switch to bypass the station optically, 2) utilizing the Secondary counter-rotating ring by reconnecting the stations electrically.

FDDI uses a timed token protocol designed to guarantee a maximum token rotation time. The timing is decided by a bidding process upon initialization of the

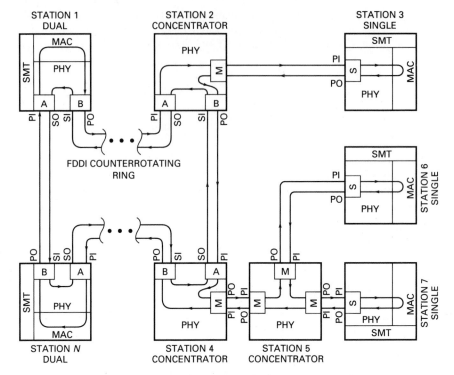

Figure 7.24   FDDI counterrotating ring interconnect example.

LAN, which permits the station requiring the fastest time between token arrivals to dictate the token rotation time for the ring. This protocol offers both synchronous and asynchronous transmissions.

The standard specifies the optical interface characteristics including the optical bypass switch, fiber optic connector, optical signal waveforms and characteristics, fiber cable type, and test methods. Some of the key characteristics are summarized below; however, the reader should always refer to the latest issue of the Standard for current information.

a) Network Performance:
   Data Rate: 100 Mb/s, 125 Mb/s clock rate
   Max Size: 100 km duplex ring (200 km fiber path), 500 stations
   Max Station Spacing: 2 km maximum between stations
   Max Station Delay: 756 nsec
   Max Delay: 756 nsec/station; 1.733 milliseconds for the loop based on 5085
      nsec/km fiber delay, 200 km loop, 500 stations
   Bit Error Rate: <10–9 BER total ring
   Tx Coding: NRZI 4B/5B

b) Media Interface Connector:

Duplex fiber optic keyed connector, ferrule in sleeve design

c) Station Optical Transmitter:

Center Wavelength: 1270 to 1380 nm

Avg. Output Power: −20 to −14 dB @ output of Test Connector

Extinction Ratio: 10% max

Spectral Width: width, plus chromatic dispersion, plus source rise time must
achieve an optical rise time less than 5 nsec in 2 km length
of fiber

Ranges from 100 to 200 nm width.

Optical Pulse: 80 nsec ± 500 ppm time interval

40 nsec ± 0.7 nsec mean pulse width

0.6 to 3.5 nsec rise-time window

d) Station Optical Receiver:

Avg Rcvd Power: −31 to −14 dBm @ input to Test Connector

Rise/Fall Time: 0.6 to 5 nsec

e) Bypass Switch:

Attenuation: 2.5 dB max

Isolation: 40 dB worst case

Time to Switch: 25 msec from command

Media Interrupt: 15 msec max

f) Fiber Plant:

Fiber Type: Multimode

Core Dia: 62.5 um per EIA 455-58

Clad Dia: 122.0 to 128.0 um per EIA 455-27, -4

Num Aperature: 0.275 per EIA 455-177

Atten @ 1300 nm: 2.5 dB/km typical, 11.0 dB max end to end per EIA 455-53

Modal Bandwidth: 500 MHz.km min@ 1300 nm optical 3 dB BW per EIA
455-30, -51, -54

Chrom Disp: see standard for profile, 0.11 $ps/nm^2/km$ between 1300 nm and
1348 nm

7.6.3 Fiber Optic LAN Optical Topology Considerations

Figure 7.10 illustrates the possible topologies for a LAN as well as for a MAN. The
three most common are the star, the bus, and the ring. Applying fiber to these topologies raises issues peculiar to the nature of fiber optics including: a) limitations as to
how many optical couplers can be placed in series on a LAN due to rapid optical
power loss through optical couplers; b) the limited ability of fiber-optic receivers to
detect signal collisions in CSMA/CD (ethernet type) LANs due to the unipolar nature
of optical signaling; c) the wide bandwidth and low attenuation of fiber, permitting
longer distances and higher data rates than copper-based LANs. This section will
concentrate on the practical limitations caused by component power losses that re-
strict fiber to the star and active ring topologies.

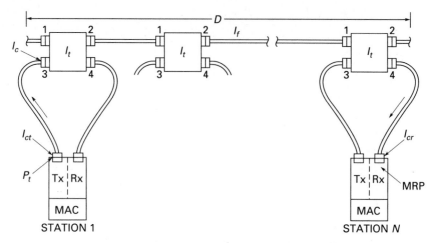

Figure 7.25   Passive fiber-optic ring or bus LAN configuration.

Figure 7.25 illustrates the physical components that make up a passive fiber optic ring and Fig. 7.26 the components that make up a passive star LAN configuration. For the ring, passive 2-port optical couplers are used to couple power from each station's transmitter onto the fiber ring and to couple common signal power from the fiber ring to the receivers at each station. The station access electronics is called a Medium Access Unit (MAU) and contains the electro/optical transmitter and receiver electronics. The MAUs and the couplers are interconnected through optical connectors.

In order to determine the number of stations or MAUs that can be connected to a LAN of this design, the designer must compute the Link Power Budget for the

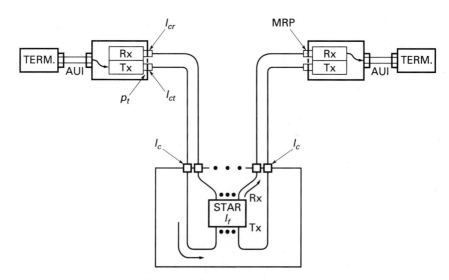

Figure 7.26   Passive fiber-optic star LAN configurations.

specific design. The optical power and loss relationships for the farthest separated stations on the LAN is found by tracing the signal from the transmitter at station 1 to the receiver at station N. For the ring configuration we get:

$$Pt - [l_{ct} + N(2 l_c) + l_{cr}] - [l_{32} + (N-2) l_{12} + l_{14}]$$
$$- D l_f - M > MRP$$

and for the star we get:

$$Pt - [l_{ct} + 2 l_c + l_{cr}] - Lstar - D l_f - M > MRP$$

where:

Pt	= transmitted power before the connector
MRP	= minimum power required at receiver after the connector
M	= assumed operating safety margin
l_{ct}	= connector loss at the transmitter
l_{cr}	= connector loss at receiver, generally low
l_c	= connector loss at couplers
l_{nm}	= loss from port n to port m of a coupler
L_{star}	= loss from port to port in STAR coupler
l_e	= excess loss in coupler
D	= longest fiber distance between stations in km
l_f	= fiber loss in dB/km
N	= number of stations

The simplified coupler loss relationship is:

Coupler loss = excess loss + power splitting loss

$$l_{nm} = l_e + 10 \log [1 / CR]$$

$$CR = P_{oc} / Po(tot) = Pic / Pi(tot)$$

$$Lstar = l_e + 10 \log(N)$$

where Pic and Poc is the power coupled at the input and output port and Pi(tot) and Po(tot) is the total input and output power.

If we assume for a 2-port coupler that $l_{32} = l_{14} = 10\log[Ptot/Pc]$, then solving for number of stations N we get the curves illustrated in Fig. 7.27. Here the number of possible stations on the LAN is plotted as a function of the quality (insertion loss) of both the optical couplers and the connectors used. Note that under most practical conditions a LAN of this design is limited to only a few (3 to 10) stations and is therefore a poor choice.

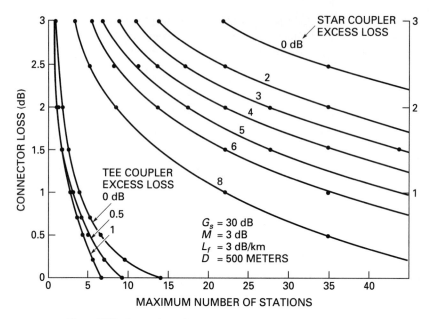

Figure 7.27   Comparison of star- and tee-coupler LAN performance [1].

REFERENCES

1. R. Hoss, "Fiber Optic Communications Design Handbook," chapter 7, Prentice Hall, 1990.

2. ANSI T1.105-1988 "Digital Hierarchy Optical Interface Rates and Formats Specification," ECSA document T1X1/87-129R1.
also
ANSI Standard for Telecommunications Digital Hierarchy Optical Interface Specifications: Single Mode, ECSA document T1X1/87-128R1.

3. Northern Telecom FD565 Ring System Applications Guide, Issue 1, Dec. 17, 1987.

4. R. Hoss, F. R. McDevitt, "Fiber Optic Video Supertrunking, FM vs Digital Transmission," NCTA, 1984.

5. R. McDevitt, R. Hoss, "Repeaterless 16km fiber Optic CATV Supertrunk Using FDM/WDM," NCTA, 1983.

6. Installation in Queens N.Y. by Warner Amex Cable, a one time joint venture between American Express and Warner Communications.

7. J. Chiddix, "Optical Fiber Supertrunking, The Time Has Come, A Performance Report on a Real-World System," NCTA, 1985.

8. R. Hoss, "Fiber Optic Options for Video Transmission," Proceedings of the Newport Conference on Fiberoptics Markets 'Where The Customers Are', Kessler Marketing Intelligence, October 22 and 23, 1985.

9. R. McDevitt, R. Hoss, "Application of Fiber Optics Networking in the CATV Industry," FOC 83, Atlantic City N.J., Information Gatekeepers.

10. Ipitek® product literature, "The Fiber Hub System," Ipitek Division of TACAN Corp., 1991, 2330 Faraday Ave, Carlsbad, California 92008, Courtesy Robert Chalfant.

11. T. V. Muoi, "CATV Supertrunking Study," by PCO Inc., unpublished, from ref. [1].

12. ANSI/IEEE Std 802.3-1985, SH09738, IEEE, 445 Hoes Lane, Piscataway N.J., 08854.

13. ANSI/IEEE Std 802.5-1985, SH09944, IEEE, 445 Hoes Lane, Piscataway N.J., 08854.

14. IEEE 802.5J-88/45, "Draft 9 on Fiber Optics," 7/88, IEEE, 445 Hoes Lane, Piscataway N.J., 08854.

15. ANSI X3T9.5/83-15 REV 15, "FDDI Physical Layer Protocol (PHY)," Draft Standard, Sept 1, 1987.

16. J. F. McCool, "The Emerging FDDI Standard," presented at the System Design and Integration Conference, Feb. 12, 1987.

17. ANSI X3 Project 503D.

18. IEEE P802-3I/D2-88-10, Twisted Pair Medium. Type 10 Base T, 7/88, IEEE, 455 Hoes Lane, Piscataway N.J., 08854.

19. M. E. Abraham, "Fiber Optic Ethernet," LAN Magazine, Nov. 1988, pp. 39–44.

20. E. G. Rawson, "The Fibernet II Ethernet Compatible Fiber Optic LAN," IEEE Journal of Lightwave Technology, Vol. LT-3, pp. 496–501, June 1985.

21. IEEE 802.6-88/58 MAN Standard Draft, D.O 6/88, IEEE, 445 Hoes Lane, Piscataway N.J., 08854.

22. ANSI X3T9.5/84-48 REV 8, X3T9 Technical Committee Draft Proposal, "FDDI Physical Layer Medium Dependent (PMD)," July 1988.

23. "Optical Fiber Video Delivery Systems of the Future," IEEE LCS, Vol. 1, No 1, February 1990.

8

Installation
and Measurements

The installation and testing of fiber-optic systems obviously calls for different procedures than those used in most electronic systems. For instance, instead of concentrating on voltage and resistance measurements, the technician will be more concerned with measuring light intensity.

In this chapter we discuss safety, installation, and fiber-optic measurements. Splicing and connectoring were discussed in Chapter 5.

8.1 SAFETY

Like any other new system, fiber-optic systems have some unexpected hazards that can cause personal injury if ignored. These dangers originate in the lasers and LEDs, the glass fibers and receiver power supplies, and in the materials and equipment used for installation and measurement. Specific precautions are given in the following paragraphs; general precautions may be found in the author's *Handbook of Electronic Safety Procedures* [1].

Chemical Hazards.   Freon and other solvents used in stripping and cleaning optical fibers may be hazardous. Avoid breathing vapors, especially if you are in a confined area such as a manhole [2]. When handling flame-retardant cables, avoid prolonged skin contact; wash hands before smoking or eating [3].

In some field connecting procedures, propanol is used as a lubricant. In such cases adequate ventilation is necessary because the fumes from the propanol are flammable [4]. Avoid contact with LOCTITE 495, an adhesive used in some cable preparation steps, as it cures instantly when in contact with human skin.

Electrical hazards.   In most circuits, only very low voltages (12 V or less) are in use. These voltages are harmless, of course, unless your feet are immersed in water at the bottom of a manhole. However, in some avalanche photodiode circuits in receivers, 300 V or more may be present. Such voltages should be treated with respect. Very high voltages are present in fusion splicers and in the oscilloscopes in optical time-domain reflectometers. Make sure that the frames of such equipment are well grounded.

Mechanical Hazards.   Glass optical fibers can puncture the skin as well as the eyes. Use protective safety glasses when handling exposed fibers and during fiber cleaving, grinding, and polishing operations.

Radiation Hazards.   It is very dangerous to stare at the end of an optical fiber that is illuminated by a laser or an LED [5]. These light sources can be hazardous whether they are part of the fiber-optic system or a part of the test equipment. Even though the radiation may be invisible in most cases, the retina of the eye can be damaged by looking at such radiation. Typical warning signs are shown in Fig. 8.1. Particular care should be taken to avoid viewing output flux under magnification.

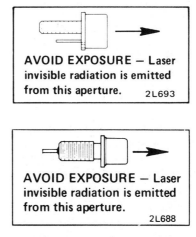

Figure 8.1 Typical laser warning labels. (From Ref. 18; courtesy of RCA.)

8.2 CABLE PREPARATION BEFORE INSTALLATION

8.2.1 Cable Design

The first rule that an installer or engineer must adhere to when planning an installation is to ensure that the cable is designed for the installation conditions, both the dynamic stress conditions of installation as well as the longer term environmental conditions that the cable will be exposed to. The cable should be ordered from the supplier based on a thorough understanding of the installation conditions and the application. As discussed in Chapter 4 (Section 4.5), shielding, strength members, gel filling, and jacket are all designed differently for different environments. The different cable designs for each installation environment are discussed in more detail in Chapter 4. A summary is given here.

Trunk cables for direct burial are different than those required for conduit or aerial. Not only are the pulling tension requirements different but the armoring and central member materials vary greatly. Because lightning can be attracted to any metal in the cable, dielectric central member materials may be required so as not to attract lightning through fibers in the event of a strike. In heavy lightning areas a copper secondary armor may be advisable to conduct the energy along the cable surface and reduce penetration. Rodents love to gnaw on the cables whether buried, aerial, or in duct. Therefore, special armoring must be used for protection against rodents. Cable designed for duct installation (and some aerial installations) may have high pulling tension requirements (600 to 1000 lb). If a lower tensile strength cable is used, fibers may break during installation or, even worse, over time due to the stress cracks that may have formed on the fiber surface.

Building distribution cable must be designed so that it can be pulled throughout buildings and up riser ducts. Since a single cable may often feed multiple floors, the cable must be designed so that fiber pairs can be conveniently broken out and terminated at different points along the cable without disturbing the remaining fibers in the cable. A cable with fibers that are individually jacketed for rough handling is generally required.

A cable is usually terminated by splicing a pigtail of fiber onto each fiber end, the pigtail containing a connector. These pigtails are essentially individually jacketed and strengthened fibers, jacketed with Kevlar* or similar strength member materials in order to prevent damage from mishandling. Sometimes bare fiber ends of a premise cable are terminated with field installed connectors. When this is done the fibers are jacketed in the field with sleeves to provide the protection or the cable is of a tight buffered fiber design so that the fibers are protected by their own thick jacketing.

8.2.2 Cable Preparation

Before installation is begun the cables should be checked to ensure that they are in proper condition for installation. This consists of testing the fiber for breaks and proper attenuation and sometimes bandwidth characteristics. Test methods are covered in Section 8.7. There are generally two inspection stages involved.

---

*Kevlar is a registered trademark of E.I. Dupont.

1. *Receiving inspection.* Where the objective is to determine that the manufacturer shipped what was ordered and to the specifications required

2. *Preinstallation tests.* Performed before the reels are sent to the site for installation to check for bent reel flanges, broken fiber, cable length, and attenuation

8.3 AERIAL INSTALLATION

8.3.1 Aerial Installation Practices

Cable for aerial installation is described in section 4.5.4.1. In the majority of aerial installations cables are lashed to a braided steel wire, called a messenger or strand, that is stretched between the poles. The construction steps are as follows [6, 7].

Make Ready. The poles must be prepared to accept the messenger that the fiber cable will be lashed to. All cables and any broken lashing wire must be checked to ensure that they are not sagging into the space that the fiber will be placed. Specifications for sag are contained in the "Safety Rules for Installation and Maintenance of Electric Supply and Communications Lines" in the *National Safety Code Handbook* [8].

Messenger Placement. Fiber cable generally does not have a messenger cable within it to carry its weight between poles. A braided steel messenger is therefore stretched between poles and connected to each pole in order to carry the weight of the cable and take any tensile forces applied by the environment off the fiber cable.

Cable Placement. The fiber cable is then hung on the messenger so that it can be lashed in place. This is generally done using one of three approaches: drive-off, back pull, or back pull with winch. These are discussed in Section 8.3.2 below and illustrated in Figs. 8.2 and 8.3.

Lashing. The lashing machine is placed on the strand and the cable positioned through it. The lashing machine is pulled with a rope and it rotates around the cable and messenger winding one or two lashing wires around the pair. Dual lashing wires are recommended.

Lashing is done span by span. When the lasher reaches a pole, a crew member temporarily clamps the lashing wire to the messenger and the lashing machine is transferred to the other side of the pole. The crew member then uses a template or free-forms an expansion loop, or swag loop [7], in the cable (Fig. 8.4). Expansion loops compensate for temperature and other loading factors which may cause stresses on the messenger and pole different from those on the fiber cable.

Once the loop is formed the lashing wire is attached to a clamp and the lashing continues down the cable. A permanent lashing wire clamp and cable support straps are installed at the loop.

Stowage. If the cable is not immediately terminated or spliced, cable ends should be looped and hung with a cable clamp, cut neatly, and capped. Cable ends for

Figure 8.2   Aerial cable placement—drive-off technique. (From Ref. 15; copyright 1986 AT&T. Reprinted with permission, all rights reserved.)

splicing should be long enough so that they can be brought down to the street to the splicing van plus have about two meters of cable removed.

Aerial Splicing.   Splicing should not be attempted aerially in a bucket. The poor environmental conditions, vibration, and fatigue will create inferior splices. Splices should always be performed in an environmentally controlled van on the ground and the splice enclosure then hung aerially on the messenger. This implies that when an aerial splice is performed a rather large service loop of cable will be required to permit the splice to be performed on the ground and then the enclosure to be hung on the pole (see Fig. 8.4).

Before splicing, the cable ends should be cut back about two meters to ensure that the gel filling has not leaked out or water has migrated in. The splice enclosure should be environmentally sealed. As with all splices it is recommended that the enclosure be re-enterable for repairs or reconfiguration.

Grounding.   The splice enclosure should have a means for grounding any metallic armor or strength members within the cable. Although grounding of fiber cable is not as critical as electrical cable, it is necessary for safety reasons and to properly conduct lightning. Grounding practices normally followed in the CATV installation industry are suggested for cable armor, center strength member, and splice enclosure housing.

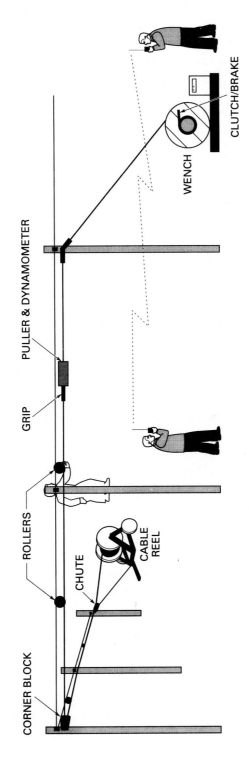

Figure 8.3 Aerial cable placement—back-pull technique using a wench.

229

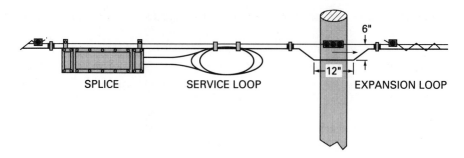

Figure 8.4 Aerial splice enclosure and expansion loop.

8.3.2 Aerial Cable Placement

Various methods are used for placing cable and attaching it to the messenger.

8.3.2.1 Drive-Off

Fig. 8.2 illustrates the drive-off approach. It is used where the route is clear such that the cable reel can be placed on a truck and paid off directly onto the pole. At the first pole, equipment called a "pusher" as well as positioners and guides (or rollers), are placed on the messenger to help hold the cable in place and guide the lasher over the cable and messenger. The lashing machine is also hung on the messenger. The guide holds the cable on the messenger ahead of the lashing machine. The truck with the cable reel is driven about 50 ft down the route and cable is reeled off and fed back through the guides, positioners, and the lashing machine. While the cable is held at the pole the truck moves slowly down the route, feeding off cable as ropes pull the cable guides and the lasher along. The lasher wraps a steel wire (or wires) around the cable and messenger. Although pulling tensions must be observed, this method places the least stress on the cable.

8.3.2.2 Back-Pull

The back-pull approach is generally used when cable must be installed over existing facilities or where the route is not as accessible by vehicles. The trailer on which the cable reel is placed is mounted on the ground about 50 ft from the beginning of the pull. Roller guides are hung along the messenger every 30 to 50 ft. A 45-degree cable chute is placed at the beginning of the pull where the cable is fed; cable blocks around bends must be of such a radius that the recommended minimum bend radius of the cable under full tension is not exceeded. Rollers are placed on all corners to maintain minimal frictional force and bend radius.

The cable reel is placed at the pull and should be braked to ensure that the reel will stop turning when the cable is not being pulled. Otherwise it may get caught in the region of the reel or sag to the ground mid span and jerk when pulling starts again.

Because pulling tensions can be high, 600 lb or more where multiple 45- or 90-degree bends exist, a pulling grip and fuse link must be installed on the cable. The pulling grip is designed so that the tensile force from the pulling rope is translated to the cable's strength member and not to the fibers or other core materials. Use a grip as recommended by the cable manufacturer. If the manufacturer has no recommended grip, often a "Chinese fingers" type of pulling grip works well but be aware that this grip translates tensile force to a combination of crushing force and tensile force on the jacket material. The end of the cable may have to be cut back a few meters when completed so no residual distortion remains. In all cases the pulling grip should be a two-piece swivel type so that twisting forces are not translated to the cable.

A fuse link, set at maximum recommended pulling tension, is recommended to prevent accidental damage to the cable. If maximum tension is exceeded it may not immediately cause fiber breakage; however, surface cracks on the fiber will be widened and the fiber may break some time later (even months later) as the cable hangs on the messenger. If a fiber does break it may break many meters down the cable, not at a point near the end. This means that the end cannot be simply cut off to eliminate the break.

A dynamometer is also recommended to be connected to the pulling rope at a point and in a manner where a crew member can read it. This is an instrument that gives a direct readout of pulling tension. Because pulling tension can be erratic when the cable binds the dynamometer cannot prevent damage; that is what the fuse link is for. The dynamometer can, however, be used to control tension and manage the pull.

Constant tension should be maintained throughout the entire pull. To maintain an even pulling force in line with the cable, a cable puller is often installed on the messenger and is attached to the pulling eye and fuse link/dynamometer assembly. As the cable puller passes, the rollers must be removed and replaced.

The cable is placed by pulling it through the initial chute and then lifting it onto the rollers that have been placed along the messenger. Cable pulling is generally done by a person on the ground or a person in a moving truck, or both depending on the section of the pull and pulling force required.

This method of installing the cable is rough on the cable, particularly where 90-degree bends are present. At all bends, corner rollers must be used that are designed for fiber cable and designed to maintain the minimum bend radius under tension.

8.3.2.3 Back-Pull with Winch

Generally where vehicle access is not possible or when there are multiple 90-degree turns which require high pulling tension, a winch can be used to pull the cable (see Fig. 8.3).

As with the back-pull approach, all cable blocks and rollers are placed on the messenger first. The winch is then set up about 50 ft from the end of the cable run. The winch line is then run along the total cable run and placed on all rollers. Special winch line blocks are used at corners so as not to damage the rubber or plastic roller blocks that are used for the fiber cable.

The winch must have a "break" on it that is set below the maximum recommended pulling tensile strength of the cable. This break releases when that tension is reached. The winch cable is attached to the fiber using a pulling grip just as in the previous case. A fuse link is still a recommended safety practice even though the winch has a break. A cable puller attached to the messenger is recommended because of the high tensile forces.

Once all is connected, the winch is tensioned and the cable pulled over the rollers to the point where lashing will start. Crews should station themselves at the reel, and at the pulling grip, and be in radio communications with the winch operator. The cable reel trailer should have some back tension on it to prevent excessive cable sag along the route.

8.4 Direct Burial Installation

Cable manufactured specifically for direct burial is required. Some of the common characteristics are described in Chapter 4 (Section 4.5.4.3).

Underground cable is generally (1) placed in a prepared trench; (2) plowed in bare or with extruded duct, or (3) placed in underground duct, either city duct or duct that is installed in a trench. This section discusses the requirements of direct plow-in.

8.4.1 Direct Plow-In

When cable is installed with a plow, the cable reel is mounted on the plow and cable is fed through the plow. In this way the trench is made by the plow at the same time the cable is inserted into the ground. A plow configuration is shown in Fig. 8.5. Adequate horsepower for soil conditions and depth must be used. Too much is better than too little.

With this method the length of cable buried at one time is dependent only on the route constrictions and the maximum length that can be ordered on a reel. Typically from 3 to 5 km lengths (and sometimes 10 km) are plowed in when right-of-way cost is not prohibitive. The installation method is typically as follows.

8.4.1.1 Route Preparation

Right-of-way is generally purchased or leased in open field or forest, along roadways or railroad tracks. The route is planned, surveyed, and documented. Stakes are placed along the route at short intervals to guide the plow.

The route plan should take account of the soil conditions and depth of burial. Rocky soil is difficult to plow in and can cause the plow to jerk and break the cable.

Figure 8.5   Direct plow-in installation. (From Ref. 15; copyright 1986 AT&T. Reprinted courtesy of AT&T, all rights reserved.)

Subsurface investigation may be required under certain conditions. The cable should be buried as deep as possible to avoid being damaged by future crossings, frost heave, or ice. Typical depth is 3 to 4 ft. Some burials at 2 ft are performed but generally only for local distribution where threat of damage is slight.

In most cases the route crosses many obstructions including roadways, streams, swamps, bridges, railroad tracks, and other buried utilities. If other utilities are crossed, the utility owner must be notified and details of the burial obtained so that the plowing can proceed without damage to those utilities. In the case of roadways, railroad tracks, and shallow streams, a rigid duct must be placed under the obstruction. For larger rivers and bridges, the duct will generally be attached to the bridge. The cable will be pulled through this duct. In some installations the cable is re-reeled after pulling through the incidental duct, and plowing continues. In others the point of obstruction becomes a splice point. For this reason careful route planning and design is needed before the burial begins.

Cable can be placed in large rivers and lakes, but special techniques must be used employing divers. Generally cable must be buried on the bottom so that dredging operations and boat anchors do not damage it.

For certain obstructions the use of other equipment may be required, such as an earth saw or boring equipment.

Where risk of cable damage is high due to frost heave, rodents, or other plowing operations, often a rigid duct is plowed in and the cable pulled through it. Typical inner diameter of the duct is 1.5 to 2 in, but it depends on the cable. Some companies can factory extrude this duct around the cable and place it on a reel. In this way the duct and cable can be buried together.

Planning is also required for placement of repeater huts. Local ordinances or the right-of-way provider may permit such buildings in only specific locations. There also must be access to power as well as road access for maintenance during all seasons. If placed along railroad tracks, the hut must be far enough away from the track so that vibration will not affect the equipment performance, and so that road-bed repair equipment can pass. Special right-of-way may be required to install the hut. Spacing is defined by the link power budget requirements of the fiber equipment as well.

A starting and finishing pit should be dug at each buried splice location and the pits should remain open during the plowing operation.

8.4.1.2 Plow Preparation

The cable feed system on the plow is important because it largely determines the tension on the cable during the plowing operation. This system consists of a reel carrier, rollers or guide tubes, and a cable chute attached behind the plow share. The feed system must be adapted such that the minimum bend radius of the fiber cable is maintained. The manufacturer generally specifies this radius under installation tension. It is usually about 20 times the cable diameter. The feed systems of these plows are generally made for large copper cable so they must be checked and if necessary modified so that rollers and such do not catch the small diameter fiber cable and pinch it.

It is particularly important that the feed at the point that the cable exists the plow share be adapted such that no sharp edges or bends are possible if the plow is abruptly jerked upwards by rocks during the plowing. A suggested design can be found in reference [9]. Some of the information in this section was derived from that reference.

The feed chute must also have a means for opening it and removing the fiber at intermediate points during plowing operations. If a vibratory plow is used, the feed chute should be isolated from the vibrating plow.

A plow with a hydraulic capstan that pulls the cable off the reel and delivers it to the feed chute produces the least tension on the cable, but conventional feed systems will work if proper precautions are taken. Tension on the cable is generally proportional to the weight of the reel and is maximum during startup or when hitting an obstruction or rough terrain. Ensure that the cable has a tensile strength of 600 lbs minimum.

8.4.1.3 Plowing

In rocky or hard soil conditions it is recommended that a ripping pass be made. This is a separate plowing operation that prepares the path of the plow and dislodges any obstructions. It is also sometimes necessary to gain the proper depth. In softer soil a ripping plow can be placed on the front of the cable plow and the operation can be done simultaneously.

To prevent cable damage, the plow should never be raised or lowered without the tractor moving forward, nor should the tractor back up with the cable chute in the ground. If the cable must be removed from a buried chute, it should be dug out.

Prepare the starting end of a cable and monitor the fiber with an OTDR during the plowing. In this way if a fiber becomes broken it will be noted at the point of occurrence and a loop of fiber left for a repair splice. If discovered later, the cable would have to be dug up and two splices made where a new cable section is inserted.

8.4.2 Trenched Cable

Cable for trench installation is identical to that for direct plow-in with the possible exception that tensile strength can be less. When PVC or steel duct is placed in the trench and the cable pulled through, the cable used is more like duct-installed cable and is generally all dielectric so as not to attract lightning.

When cable is direct buried in a trench, the trench is first dug, then prepared; the cable is then placed in the trench and the trench refilled. Trenching is often done where either route or cable conditions prevent plowing. Trenching is the most gentle means of placing underground cable. Tension is only applied when laying the cable into the trench from the reel.

Once the trench is opened all sharp rocks should be removed from it before laying the cable. If it is in a rocky area then the trench should be filled with at least 6 in of select fill before cable placement. If the soil that was dug out is to go back in, all sharp rocks should be removed from it, or if not possible, at least 6 in of select fill should be placed over the cable.

When laying the cable in the trench it should be done so that the tension and bend radius of the cable are not exceeded. When duct is installed in the trench, pull boxes are placed periodically and pulling ropes are installed on the duct. Cable is pulled, sometimes manually, from pull box to pull box.

8.4.3 Duct Installation

Cable manufactured to withstand the pulling tensions, the crush, and the environmental conditions of duct installation is required. This is described in Chapter 4 (Section 4.5.4.2).

During installation fiber breakage can occur many meters inside the end of the cable; therefore, adequate excess must remain at cable ends so that it can be cut off to expose good clean continuous fiber.

When cable is installed in duct the cable reel is mounted on a carrier on the street or a truck. The configuration is shown in Fig. 8.6. With this method the length of cable pulled at one time is dependent principally on the duct route, duct conditions, and the number of turns and manholes. Typically from 3 to 5 km lengths are able to

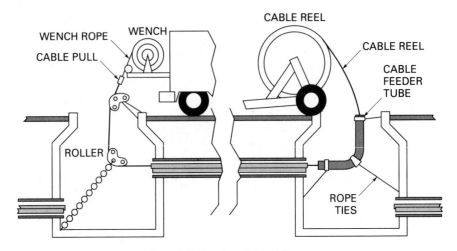

Figure 8.6   Duct installation of fiber.

be pulled if cable looping and payout techniques ("figure 8" approaches, for example) are employed at manholes. The installation method is typically as follows.

8.4.3.1 Route Design

Right-of-way is generally purchased or leased from utilities or the municipality. The route is planned and the duct maps and duct inspected to determine availability. Final assignment comes after duct rodding and roping, which determines whether the assigned duct is clear and in good shape.

The route should be planned and inspected by both the outside plant engineer as well as the construction supervisor. In addition to the right-of-way constraints, and route plan should take account of any access points to buildings that are to be entered and the need and cost of any street cuts. Water, poisonous gas, and hot steam conditions should be considered as well. If steam tunnels are close to the duct route, cable design must account for the long-term heat effects. If cable is to be under water, both the jacketing and gel filling must accommodate this.

When fiber is installed in duct it is generally within a city and distances are such that repeaters are not required. If they are required, however, repeaters should be installed within buildings, never within standard manholes. Splices, conversely, can be installed within manholes if special precautions are taken. These precautions will be discussed in later paragraphs.

Planning is required for placement of cable. The route and placement plan must consider the cable reel lengths, so as to maximize cable use and minimize the number of splices. Reels must be numbered and identified as to placement on the route before pulling.

8.4.3.2 Rod and Rope

Before cable can be pulled through duct, the duct must be clear and a pulling rope placed in it. Clearing the duct is performed by pushing through rods or other devices to clean out all old steel rods and other material that may be in them and checking to ensure they are not collapsed. Generally nylon pulling rope is placed in the duct by blowing it through on the end of a parachute-like device.

8.4.3.3 Subduct Placement

If possible it is best to simply pull the cable through the cleaned duct. In some cases, however, the duct is large enough that multiple fibers can be pulled over time and route owners would like to retain the option of doing so. In this case subduct is pulled into the duct using the pulling rope installed. From two to three subducts are generally pulled simultaneously. The subduct typically has a 1.5-in inner diameter. Each subduct has a pulling rope installed in it. Corrugated subduct is often used because of the lower surface friction it offers.

8.4.3.4 Pulling Preparation

The cable feed setup is important because it largely determines the tension on the cable during the pulling operation. This consists of a reel carrier, rollers or guide tubes, and a cable chute attached at the manhole where cable enters and exits (see Fig. 8.6). The feed system must be constructed such that the minimum bend radius of the fiber cable is maintained. The manufacturer generally specifies this radius under installation tension. It is usually about 20 times the cable diameter.

Tension on the cable is a function of the length of pull, friction of the duct or subduct, number of turns, and to the weight of the reel. Ensure that the cable tensile strength is either 600 or 1000 lb depending on conditions.

8.4.3.5 Pulling

Prepare the starting end of a cable and monitor the fiber with an OTDR during the pull. In this way if a fiber becomes broken it will be noted at the point of occurrence and the fiber section can be removed and spliced at the nearest manhole. If discovered later, the cable section would have to be cut and pulled out and two splices made where a new cable section is pulled back in.

The cables can be pulled continuously through multiple manholes. Construction techniques such as "figure 8" methods can be used to achieve continuous pull without cutting the cable on the reel or overstressing the cable.

Start by placing the cable reel trailer and pulling winch next to the manholes. Attach cable rollers and feeders and winch cable blocks in positions above and within the manhole so that the winch line and cable will roll freely into and out of the manhole and duct (see Fig. 8.6). Rollers and feeders must be designed to maintain minimum bend radius of cable under installation conditions.

Cables should not be pulled around bends within manholes unless rollers are placed in the manholes to reduce pulling tensions. It is best to pull cable from manhole to manhole, removing the cable at each manhole and using a "figure 8" configuration to organize it on the ground in preparation for the next section pull.

Because pulling tensions can be high, a pulling grip must be installed on the cable just as described by aerial cable in Section 8.3.2. If maximum tension is exceeded it may not immediately cause fiber breakage, rather surface cracks on the fiber will be widened and the fiber may break some time later and at some length down the cable.

Once all cable blocks and rollers are placed on and in the manholes, a motorized winch is set up at the exit manhole end of the cable run (Fig. 8.7). By using the pulling ropes that were placed in the duct, the winch line is pulled along the total cable run and placed on all rollers.

Crews should station themselves at the reel and at all manholes involved in the pull, and should be in radio communications with the winch operator. The winch must have a "break" on it that is set below the maximum recommended pulling tensile strength of the cable. This break releases when that tension is reached. The winch cable is attached to the fiber using a pulling grip.

Figure 8.7   Duct installation—Capstan winch in operation. (From Ref. 16; copyright 1986 AT&T. All rights reserved, reprinted with permission.)

As the cable comes off the reel apply a cable lubricant pulling compound (one that is compatible with the cable jacket) to reduce friction.

8.5 Premise Cabling

The installation of fiber optics in buildings usually involves the same precautions and routing arrangements as copper. Fig. 8.8 illustrates a typical premise installation, showing the various configurations that fiber may be subject to.

Vertical installation requires riser cable installed within riser conduit or within riser shafts to telephone closets on various floors. Vertical distribution requires cable installed between these telephone closets and the workstations or terminating electronics on the floor. If the fiber extends outside the building the riser cable must either be interfaced (connected or spliced) with the outside plant cable or the outside plant cable must double in part as riser cable. Either choice has consequences in cable and installation design.

In a common case (illustrated in Fig. 8.9) the outside plant cable will extend into the building and be connected or spliced with specific fibers in the riser cable that connects outside services to terminal equipment in a telephone equipment room. Some codes require that gel-filled non–fire-retardant outside plant cable extend no more than a specified distance (for example, 50 feet) into the building.

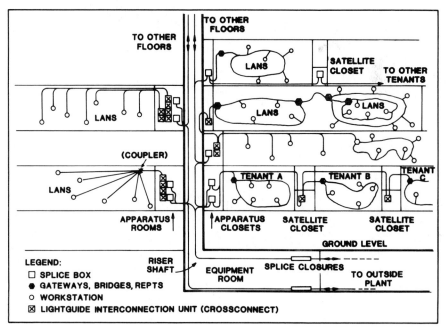

Figure 8.8   Premise cabling configuration. (From Ref. 13; copyright 1985 AT&T. All rights reserved, reprinted with permission.)

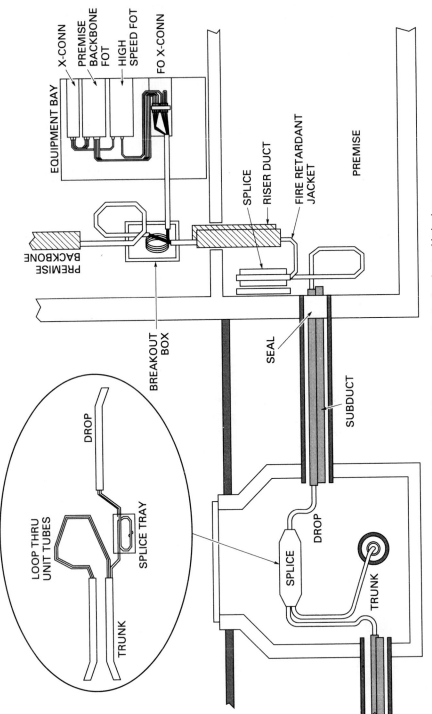

Figure 8.9 Building entry approach using breakout cable in riser.

Fibers in the riser may serve inter-floor communications. Because intra-building and external communications may differ, the riser cable may have mixed fiber types within it (singlemode for high-speed external, multimode for low-speed internal) or a separate cable may be used for the two. Riser cable will often have a different type jacketing and fill than outside plant, as dictated by codes.

Typically the riser cable is terminated at the equipment rooms by splicing connector pigtails to the active fibers. The connectorized pigtails may then terminate directly on the fiber terminal equipment or on an optical patch panel. The choice of a patch panel depends on whether the optical termination must be flexible to frequent change or not. When a patch panel is used, jumpers (fiber with connectors on both ends) connect to the fiber terminal.

The termination to the cable is achieved either by splicing pigtail connectors onto the cable or by field assembling connectors onto the fiber ends of the cable.

8.5.1 Cable for Premise Installation

8.5.1.1. Building Entry Cables

Building entry cable enters the premise from a serving manhole or feed point outside the building. This entry cable may be of a common design to outside plant trunk, or alternatively a special entry cable spliced at the feed point or manhole. In any case it generally has the same design as outside plant trunk cable since the environmental conditions are similar.

Entry cable contains multiple singlemode fibers (although some multimode is common) with fiber counts typically from 12 to 64 fibers. Entry cable usually features heavy duty armoring and sheaths, and fibers surrounded by gel filling to prevent water penetration, much as outside plant cable does (see Chapter 4).

If the entry cable can be spliced to the outside plant trunk cable, outside of the building, it may be advisable to use a fire-retardant sheath design to enter the building. Outside plant cable that is not fire retardant should not be used inside a building or beyond 50 ft of the building entrance [13]. Some codes require this; others permit non-fire retardant cable if it is enclosed in duct.

8.5.1.2 Riser Cable

Riser cable contains multiple fibers but the structure of cable jacket, strength member, and even the fiber buffering is modified to match the requirements of the building installation. See Chapter 4 (Section 4.5.5).

The outside sheath may be required by code to be fire retardant although some codes will permit flammable sheaths if installed in duct. Check before ordering.

Riser cable must support its own weight in lengths of about 500 ft without adverse effects on performance, and should also have a reasonable bend radius requirement so that it can be pulled within interior riser duct and shafts.

Generally the fibers within this cable are broken out of the cable at multiple points up the riser. The cable core construction and fiber units must be designed for this purpose.

8.5.1.3 Building Cable

Building cable runs between telephone closets (optical patch panels) and terminal equipment (LAN stations for example) across the floor. It can be run in an individual star drop configuration or can loop around as a distribution cable and have fiber pairs dropped at equipment locations.

Being in open areas it is generally required to have a fire retardant outer jacket. The cable should be purchased with color-coded individually buffered fibers so that the fibers can be fanned out of the cable and individually connectorized. Fiber count is optional but generally is ordered with 4 to 12 fibers.

8.5.1.4 Pigtails and Jumper Cords

Pigtails and jumper cords are used to connect communications equipment or terminate cable to communications equipment at distances of 100 ft or less. These cords or pigtails contain generally one or two fibers. They are heavily buffered with fire-retardant PVC (or other materials as code requires) and often contain a dielectric strength member (such as Kevlar) so that they can be handled, routed through equipment bays, and terminated within splice enclosures and optical connector bays. They are not designed to be pulled through ducts or shafts.

Pigtails have a connector on one end and a bare fiber on the other. They are spliced to entry or riser cables in order to terminate these cables, in lieu of field connectoring the cable. Splicing of a connectorized pigtail on an entry cable is usually easier than placing a connector on the cable, and less prone to accidental damage.

Jumpers have the same design as pigtails but have a connector on *both* ends. Jumpers are used for optical connections between connectors on an optical patch panel or between optical patch panels and terminal equipment.

8.5.2 Installing Premise Cable

Optical cable installed in vertical runs is required by the National Electrical Code to have fire-resistant characteristics such that it will prevent the carrying of the fire from floor to floor. If not, the cable must be encased in a noncombustible conduit or be located in a fireproof shaft containing fire stops at each floor (refer to section 770-6 of the National Electrical Code also reference [14]).

Riser shafts are designed differently in various buildings, so the options within the code to be followed are somewhat building dependent. Fiber ordering should reflect individual building differences. Riser shafts generally run straight between basement and the top floor but duct leading to the shaft may have 90-degree bends, particularly in the basement areas.

Fiber cable is generally installed in vertical risers by providing support every few floors. This can be done by using a cable grip that holds the cable sheath. The cable grip is strapped to convenient support beams, makeshift rods, or wall straps as illustrated in Fig. 8.10. AT&T recommends support every third floor or 35 ft.

AT&T suggests the following formula for determining cable length to be ordered for riser [13]:

Length = building height + (15 feet × no. closets) + 10% slack allowance

Unless circumstances dictate otherwise, cable is unreeled and installed down the riser shaft from the topmost closet. The slack should be left at the bottom. Cable loops about 15 ft in length should be formed in each closet and secured by pulling the cable back up into the closets. The jacket can then be stripped from a portion of the service loop (about 2 m) and the fibers pulled out for splicing or connectoring at the patch panel in the closet.

Termination is done after cable installation. Trying to install connectorized cable is not only difficult but the risk of damage to the connectors is too great.

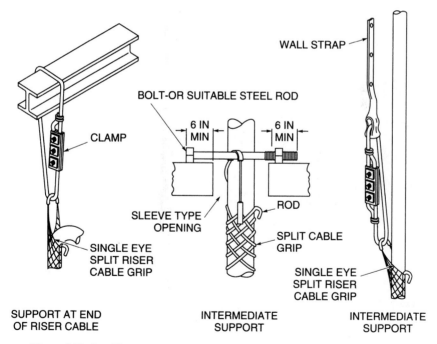

Figure 8.10   Installing cable in vertical riser. (Reprinted courtesy of AT&T, all rights reserved.)

8.6 SPLICING PROCESS

8.6.1 Splicing Locations

Fiber cable is generally spliced at the following points along a plant route.

8.6.1.1 Trunk through Splice

This is performed where two sections of trunk cables are spliced either because of finite reel lengths or because installation constraints limit the length of a cable span. If the cable is installed underground in city conduit, the splice is placed in a manhole. If the cable is direct buried then the splice is placed in a vault. The vault is generally a prefabricated concrete or composite box (often with no bottom so water can exit) that contains an access lid. If the plant is installed aerial then the splice is placed on the strand as indicated in previous sections.

The spliced fiber is always protected by overcoating with RTV (or similar soft protective substance) and secured within a protective organizer (often called a splice tray), which is then placed within an environmentally sealed enclosure such as that illustrated in Fig. 8.11. For outside plant the enclosure should be re-enterable for future repairs as well as for future changes.

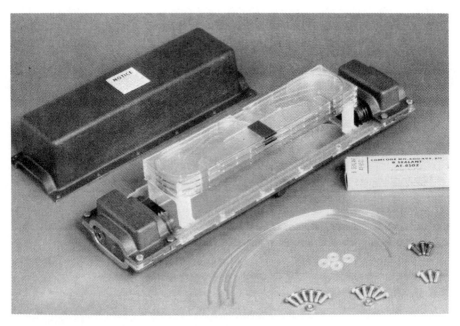

Figure 8.11 Splice enclosure with environmental seal for outside plant environment. (From Ref. 16; courtesy of AT&T. All rights reserved.)

8.6.1.2 Building Drop Points

Where a building requires access to the trunk plant, a drop or feeder cable from the building must be spliced into some of the fibers in the trunk cable. At this point the trunk cable can be brought into the building and spliced inside, or the splice may be performed outside, in a manhole, vault, or pedestal. For most public systems the splice is performed outside so that the trunk is not involved with the building.

At the splice point, it is common for all three cable ends to be cut and the fibers arranged within the enclosure and spliced to form the through and drop connections. An alternative approach, which requires much less splicing, is to cut only those fibers in the trunk cable that need to be spliced to the drop. This requires a special technique, special tools, and a cable that will permit the fibers to be pulled out individually or in groups. The fibers that continue through are placed aside, unbroken, in one set of splice trays. The fibers to be spliced with the building drop cable are spliced together in separate splice trays. The entire assembly is installed into a single enclosure with the three cables extending from it.

8.6.1.3 Cable Terminations with a Building

When cables enter the location where they are to be terminated with the electronics or on a patch panel, they must be terminated with connectors. Connectors kits are available that will permit the connectors to be terminated on the ends of the cable. This approach can be labor intensive and prone to poor yields; sometimes the connector quality is not as good as factory connections. It is often better to splice pigtail fibers that have factory assembled connectors on them, onto the cable end.

The splicing is performed with the same technology as used with outside plant splices, and the splices are protected and secured in splice trays in the same manner. The difference is that the enclosure does not have to be environmentally sealed. It is usually an aluminum box like structure that secures the cable ends and holds the splice trays and may even contain a connector mounting panel. It can be obtained in wall-mounted or rack-mounted configurations. Figure 8.12 illustrates a rack-mounted variety.

8.6.2 Splicing Technology

Chapter 5 (Section 5.1) discusses the most common splicing technologies, fusion and mechanical, and the procedures involved in the fiber splicing process.

8.6.3 Fiber Protection Precautions

Whatever approach is used for splicing, quality control measures must be strictly adhered to in the cable-end preparation and subsequent protection and stowage of the fiber. Strict measures must be taken such that bare glass is protected in the splicing

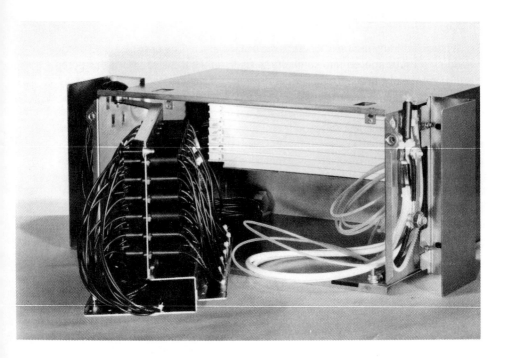

Figure 8.12   Rack mounted splice shelf and enclosure including optical patch panel. (From Ref. 17; courtesy of Northern Telecom.)

process from dirt and direct handling. If the bare fiber is scratched by dirt or dust particles, a small surface defect will remain in the glass. When this surface defect is subject to tension it will grow and propagate through the glass, eventually resulting in a break. The process is known as stress corrosion. The tension on the fiber that brings about the crack growth can simply come from the looping of the fiber within the splice tray or from dimensional changes within the enclosure due to temperature changes. Sometimes splices will break months after they have been completed.

If cable has been sitting in water, the ends must be cut back a few meters and the core inspected to ensure that the fiber has not been exposed to water. Water aids the process of stress corrosion and deteriorates the coating on the fiber. All reel ends should be cut back a few meters and inspected to ensure that the gel filling has not escaped and left voids.

When fibers are stowed in splice trays after splicing, bare fiber must be 100% overcoated for protection with an RTV or similar soft compound to prevent exposure of the bare glass. Sandwich or shrink tubing type protective fixtures are sometimes used but these approaches must be used with caution as they are more prone to collecting dirt and pressing it into the bare fiber.

8.6.4 Splice Enclosures

Splice enclosures consist of splice organizers and a protective enclosure. Fig. 8.11 illustrates an enclosure with the organizers mounted on it. In this case the organizers are in the form of trays that fan out like the pages of a book so that groups of spliced fibers can be handled separately.

8.6.4.1 Splice Organizers

Splice organizers (or splice trays), the clear plastic trays illustrated in Fig. 8.11, perform the following functions

1. The organizer secures and sometimes protects the fiber splice section. The organizer will often have a grooved section mounted in it that holds individual spliced fibers. The grooves not only hold the fibers but the RTV coating as well.
2. The organizer is shaped so as to hold the service loop of buffered fiber on either side of the splice and ensure it does not tangle with other fibers. The tray is large enough to maintain the minimum bend radius of the fiber. It is common for organizers to hold the number of fibers and splices so as to match the number of fibers in a cable core subunit. Six to 24 fibers is a common number per tray, depending on manufacturer.
3. The organizer is also designed to secure the fiber unit tubes if the cable is a loose tube design.
4. A properly designed organizer contains guides that maintain the fiber in a proper configuration and at a proper bend radius with no risk of being pinched when the organizer is secured in the enclosure.
5. A properly designed organizer and enclosure permit the stacking of multiple organizers within a single enclosure, and permit easy access to any organizer in the stack. With proper design, splices, once stowed, can be easily retrievable for rework without pulling up or breaking adjacent splices.

8.6.4.2 Splice Enclosure—Outside Plant

For outside plant application the enclosure is an environmentally sealed, stand-alone unit that is mounted within the cable right-of-way on poles, strand, manholes, vaults, pedestals, or buildings. It has the following features or characteristics:

1. The enclosure is rugged and resistant to tensile, puncture, bending, corrosive, abrasive, and other environmental forces.
2. The enclosure contains brackets for attachment to strand, manhole fixtures, walls, ceilings, or other fixed attachment areas in the cable right-of-way.

3. The enclosure holds the splice trays in proper position relative to the cable entry and service loops.

4. It secures the cable by providing clamp or tie-off points for the cable strength member.

5. When the cable requires grounding, the enclosure provides grounding clamps.

6. The enclosure (Fig. 8.11) contains rubber grommets (usually dual) that fit around the cable jacket as seals. It also contains grommets that seal the enclosure. Although not shown in Fig. 8.11, for underground environments and in manholes where water is present, an external plastic housing may also be placed over the inner enclosure and a rubbery sealant placed in the air space between the two. Because the sealant has the consistency of RTV or a gum rubber, the enclosure can be re-entered if necessary.

7. In the underground environment where the enclosure is often immersed in water, a second outer enclosure may be required. This is generally a plastic enclosure that goes over the inner one, and holds a sealant compound that maintains the water seal. Because the enclosure must be re-enterable, the sealant is generally of a type that can be peeled away upon entry.

8.6.4.3 Splice Enclosure—Equipment Room

Where riser cable is routed to the equipment room, the splice is used either for the termination of connector pigtails to the riser cable or for the connection of outside plant cable to riser cable.

In the case where it is an outside plant to riser cable splice, the option is to either use an outside plant enclosure or an interior equipment enclosure. In either case the enclosure is generally wall mounted, either at the point of building entry or closely thereby.

In the case where it is a splice of a cable to connectorized pigtail, the enclosure is generally mounted with the equipment rack or wall mounted adjacent to it. In this case it is not environmentally sealed. It often contains both the splice enclosure and a connector cross-connect as illustrated in Fig. 8.12. It has the following features or characteristics:

1. The enclosure is typically an aluminum case with front or rear panel access.

2. The enclosure may contain optional configurations depending on wall or rack mount.

3. The enclosure holds the splice trays, and contains attachment areas for cable and space for stowing service loops of fiber.

4. It secures the cable by providing tie-off points for the cable strength member.

5. The enclosure often includes a connector patch panel as well so that the complete interconnection between entry cable and equipment connectors can be made within the enclosure.

8.7 TESTING AND OPTICAL MEASUREMENTS

During and at completion of construction of the fiber plant, quality control and acceptance testing is performed to ensure that splice loss and cable performance is within required parameters. Testing can be divided into three categories: receiving inspection, construction testing, and acceptance testing. In rare cases a receiving inspection operation can contain fiber bandwidth, dimensional, and environmental inspection systems.

Equipment also goes through receiving inspection and final acceptance testing. Receiving inspection is generally only visual. Acceptance testing is to ensure all the equipment parameters and functions are within specified parameters and that the system operates over the full span distance. In addition to a functional and optical power test, a BER test is performed.

Test equipment required usually consists of an optical time domain reflectometer, an attenuation test set, an optical power meter, and an optical attenuator. These instruments are discussed in the following sections.

8.7.1 Principle of Optical-Power Measurements*

Measurements are just as essential in the maintenance of fiber-optic systems as they are in strictly electronic systems. However, it is sometimes difficult to relate measurements by two different companies or organizations. Only when the test setups are identical or conform to a standard, can there by any assurance that test measurements are comparable.

Conventional measuring techniques can be followed in adjusting and troubleshooting the strictly electronic portions—receiver and transmitter—of a fiber-optic system. Therefore, we consider only *optical* measurements in this text.

Optical measurements require special test equipment, but simple techniques can be used in an emergency. For example, if a fiber-optic cable has a metallic shield, shorts to ground can be checked with an ohmmeter. If there is a short to ground, this may indicate cable damage, which may imply fiber damage. Another simple test on short spans is to shine a flashlight into a fiber at one end and then look for light at the other end [7]. (*Careful:* Use of the eye as a sensor could lead to the bad habit of using the eye in all cases, even in hazardous situations.)

Specific tests performed during installation or maintenance are discussed later. The most basic optical power measurements are generally the following:

1. *Absolute radiant power output of the source,* which is as important to fiber optics as absolute current and voltage are to electronics. If an optical source delivers significantly less than its rated output, lowering the total dB loss of a system's passive components will not compensate for the weak input.

2. *Fiber power loss,* which depends on the fiber length and on the angle of launch; measurement of fiber loss can be difficult.

*Parts of this section is reprinted with permission from Ref. 20 (*Electronic Design,* Vol. 27, No. 21); copyright Hayden Publishing Co., Inc., 1979.

3. *Connector and splice losses,* which involve specifications of both the light coupled into a fiber's core and the light coupled into its cladding; fiber specs are usually for light coupled into the core only. The surest resolution to this inconsistency is to measure the optical power into and out of the connector or splice.

4. *Receiver sensitivity,* which determines the noise performance of the span. Whether or not the system photodetectors provide gain, they convert incident light into electrical current. Measuring the efficiency of the conversion—or responsivity—requires an optical-power meter and an ammeter.

It is most convenient to measure all these variables directly in decibels (dB)—the standard communications unit. When expressed in dB, system gains and losses can be easily perceived and quickly evaluated.

Absolute-power measurements in dBm and dBµ are important for evaluating active optical components, such as sources and receivers. Measurements of relative-dB loss are appropriate for passive optical components, such as fibers, connectors, splices, and tees. Some popular optical dB units are described in Table 8.1.

TABLE 8.1 ENLIGHTMENT ON OPTICAL dB UNITS

Measurements of optical radiation power are expressed in watts. Decibel (dB) power is

$$dB = 10 \log \left(\frac{P_{sig}}{P_{ref}} \right)$$

where P_{sig} is the power to be measured and P_{ref} is the reference power.
For 1-mW reference power

$$dB_m = 10 \log \left(\frac{P_{sig}}{1 \text{ mW}} \right)$$

For 1-µW reference power

$$dB_\mu = 10 \log \left(\frac{P_{sig}}{1 \text{ µW}} \right)$$

With both P_{sig} and P_{ref} variable, the dB-power formula expresses the log ratio of the two unknowns in dB. Light-power loss of an optical data-link element is

$$L \text{ (dB)} = 10 \log \left(\frac{P_o}{P_m} \right)$$

The inner power (P_{in}) and output power (P_o) of a component can be measured in dBm units, dBµ units, or dB units without a known reference. The loss, L, expressed in decibels, is the same:

$$L \text{ (dB)} = dBm(P_o) - dBm (P_{in})$$
$$= dB\mu (P_o) - dB\mu (P_{in})$$

Source: Reprinted with permission from Ref. 20 (*Electronic Design,* Vol. 27, No. 21); copyright Hayden Publishing Co., Inc., 1979.

LEDs and injection laser diodes, the most popular fiber-optic emitters, have small active areas; their total light-output power, therefore, is readily measurable by a large-area photodiode detector placed nearby, as the setup in Fig. 8.13 shows.

Photodiode sensors with areas of only 1 cm^2 can handle half-cone acceptance angles (θ_m) greater than 45 degrees. This acceptance angle corresponds to a numerical aperture for light collection (sin θ_m) of more than 0.7, which, of course, is quite high.

The numerical aperture is also a measure of the angular distribution of power emitted from the source. This distribution can be tested by scanning a pinhole mask across the detector's field of view of the source.

Sources for fiber-optic communications systems usually emit in the spectral range 700 to 144 nm. The spectral response of silicon photodiodes matches the range from 700 to 1100 nm; germanium devices can measure sources from 1000 up to 1400 nm.

Silicon detectors are more sensitive than germanium detectors and can measure lower power. A 1-cm^2 silicon photodiode can measure absolute power down to 1 pW. Higher leakage currents and lower internal impedances limit germanium diodes to 10-nW measurements. However, germanium diodes, unlike silicon photodiodes, can measure wavelengths longer than 1100 nm.

The power output of the source is one of the factors involved in the calculation of fiber loss, which is most conveniently represented by the cable-loss factor (CLF) in dB/km:

$$\text{CLF} = \frac{P_i - P_o}{L}$$

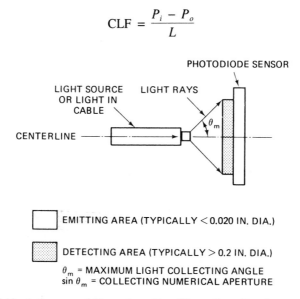

EMITTING AREA (TYPICALLY < 0.020 IN. DIA.)

DETECTING AREA (TYPICALLY > 0.2 IN. DIA.)

θ_m = MAXIMUM LIGHT COLLECTING ANGLE
sin θ_m = COLLECTING NUMERICAL APERTURE

Figure 8.13 Active areas of fiber-optic emitters. Fiber-optic emitters have small active areas, so their total light-output power is readily measurable with a large-area detector placed near the source. (Reprinted with permission from Ref. 20 [*Electronic Design*, Vol. 27, No. 21]; copyright Hayden Publishing Co., Inc., 1979).

where

$$P_i = \text{input power, dB}$$

$$P_o = \text{output power, dB}$$

$$L = \text{cable length, km}$$

The power output of the source is easier to measure than the effective input power to the fiber. Loss mechanisms extract a heavy toll on light sent to the fiber's core; for example, not all of the source light falls on the core and the launch angle may direct some of the light into the fiber cladding.

The difference between the optical power input and the optical power output of a connector or splice equals either the connector loss or the splice loss. Making measurements directly in dB simplifies the required calculations. If the connector in question is attached to a long cable, the CLF must be subtracted from the total loss measured for the connector.

The overall measure of power losses from a system may not correlate to the manufacturer's specifications for losses from individual components. Some makers of connectors specify the loss both for light coupled into the fiber core and for light coupled into the cladding. However, cable specifications often include only the light coupled into the core. In addition to the losses specified for connectors and cable, an accurate appraisal of total-link loss requires measurement of the loss into fiber cladding at a connector interface.

The quality of the conversion of input optical power into output electrical current is called responsivity (measured in A/W). To make this conversion, photoreceivers for fiber optics most frequently employ photodiodes (composed of Si, GaAs, or Ge) as the basic photosensor. Photodiode signals are amplified and processed according to the requirements of the system. The photodiodes can be PIN types, without internal gain, or avalanche types, with internal gain. The responsivity of the photodiodes is usually linear but varies with wavelength, which must therefore be measured at the source.

Both an optical power meter and a current meter are needed to measure photodiode responsivity. First, the optical power emitted from a stable source at the desired system wavelength is measured with the power meter. The calibrated source excites the current from the photodiode. This current is then measured by the current meter. The ratio of optical power to excited current is the photodiode responsivity.

The photodiode responsivity (A/W) and the conversion factor (V/A) for the amplifier following the photodiode equal the overall receiver transfer function (in volts out per watt of optical power in). The receiver transfer function, together with the output noise voltage of the photoreceiver (measured with the optical power off), determines the system's minimum ability to detect light.

8.7.2 Optical-Test Equipment

8.7.2.1 Optical-Power Meter

A typical optical-power meter (Fig. 8.14) consists of a calibrated pho-todetector that has a large area so as to capture all light entering the aperture. The detector may also perform wavelength calibration. It is generally used to determine the transmit power from an optical transmitter, test source, or from a transmitting fiber pigtail.

Figure 8.14   RIFOCS Model 555A optical power meter (courtesy of RIFOCS Corporation, Camarillo, California). The optical power meter is illustrated with a companion set of precision DIAMOND Universal Connectors, which permits use with multimode and singlemode fiber and provides interface compatibility with all industry standard fiber optic connectors. The unit provides optical power measurement, as well as attenuation and insertion loss measurement, over the 850 to 1550 nm range with 0.01 dB resolution.

8.7.2.2 Optical-Time-Domain Reflectometer (OTDR)

Fig. 8.15 illustrates the principle of operation of an OTDR. A short (nanosecond) burst of laser light is coupled into the fiber being measured, using an optical directional coupler. This coupler permits light to pass from port 1 to 2 in the transmit direction, but light traveling in the opposite direction will be coupled into port 3 where it is absorbed by a high speed optical detector. The OTDR plug-in or wavelength setting (dual wavelength systems) is selected based on the operational wavelength of the fiber.

As the coupled light travels down the fiber some of it is scattered back owing to the normal scattering behavior of light traveling in the glass fiber. At the opposite end of the fiber a lot of the light is reflected back due to the refractive index difference between the glass and the air, in effect forming a mirror. The reflected and back-scattered light is absorbed by other loss mechanisms within the fiber span, including splices and connectors. By measuring the reduction in light from the time the pulse is emitted to the time the end reflection is reached, we can measure loss per unit length of fiber.

By recording the time between emitted pulse and the reflections from the splices, connectors, and end-face, and by calibrating for the refractive index of the fiber, we can measure distance to these items.

Fig. 8.16 illustrates an OTDR from SIECOR Corporation and Table 8.2 describes the features and specifications of both the singlemode and multimode modules.

Although the OTDR can be fairly accurate on overall loss and distance measurements, it can be misleading as to individual splice or connector loss. For example, a splice that has a high reflective property in the transmit direction but has very low loss in the other, may show up as a net gain in power instead of a loss. Likewise transitioning from a low scattering loss fiber to a high scattering loss fiber may show up as a gain in power at the transition point. Fig. 8.17 illustrates the histogram of

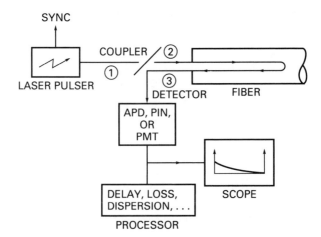

Figure 8.15 Operating principle of an OTDR. (From Ref. 21; copyright 1978 IEEE.)

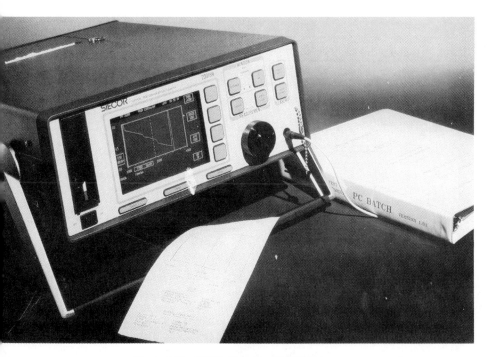

Figure 8.16   OTDR. (Courtesy of SIECOR.)

series of AT&T rotary splices measured with an absolute loss measurement test set as well as the same series of splices measured with an OTDR [22]. Although the average loss was nearly the same, the variation in measurements across the sample with the OTDR illustrates the relative inaccuracy when measuring individual splices.

8.7.2.3 Attenuation Test Set

An optical test set is illustrated in Fig. 8.18 [15]. It contains both a calibrated and stabilized optical source (or sources of various wavelengths) and a calibrated optical detector. The optical cable is connected to the test set by using a test jumper of known optical loss. A block diagram of a test set from AEG-Telefunken is illustrated in Fig. 8.19.

8.7.2.4 Attenuator

An optical attenuator is used to insert optical loss into the transmit to receive span. Fixed attenuation is inserted in order to set the proper operating range of the equipment when an excess power condition exists. A variable attenuator is used during acceptance testing to determine the amount of optical power margin that exists in a span, by increasing loss until performance decreases to the limits specified.

Fixed attenuators are inserted at an optical cross-connect or at the terminal

TABLE 8.2 SIECOR 2001HR MULTIMODE AND SINGLEMODE OTDR SPECIFICATIONS

Single-mode module specifications

Optical and display specifications

Single-mode plug-in modules	2001=1300S	2001=DUALS	
Wavelength (nm)	1310 = 20	1310/1550 = 20	
		1300 nm	1550 nm
Pulse width (ns) at 1300 nm	20/75/250/	20/75/250/	40/100/250/1000/6000
	1000/6000	1000/6000	
Measurement range backscatter (dB)	3/7/9/14/18	3/7/9/14/18	3/5/8/13/16
Dead zone (attenuation)	≤ 15 m	≤ 15 m	≤ 20 m
Dead zone (event)	≤ 5 m	≤ 5m	≤ 10 m
dB readout resolution	0.01dB		
Distance range	0 to 100		
Data acquisition window (m or ft)	Selectable over full distance range		
Refractive index range	1.400 to 1.700		
Distance measurement accuracy	= 0.01% = 0.5 m at 10 km		
	= 0.01% = 5.0 m at 100 km		
Minimum length readout resolution	0.02 m or 1[a]		
Horizontal scales	3 m (smallest window) to 100 km (largest window)		
Averaging time	Selectable (real time, 30 sec., 2 min., continuous)		
Display	6[a] diagonal screen, green		
Printer output	Integrated high-speed graphics printer (optional		
Connector types	Biconic, FC, D4, SMA, DIN, SC, ST compatible		

Multimode module specifications

Optical and display specifications

Multimode plug-in modules	2001=0850M	2001=1300M	2001=DUALM
Wavelength (nm)	850 ± 20	1300 ± 20	850/1300 ± 20
Pulse width (ns)	5/50/100/200	5/50/200/1000	same as single wavelengths
Measurement range backscatter (dB)	5/10/13/16	3/6/12/15	4/9/12/16–3/5/11/15
Dead zone (attenuation)	≤ 15m	≤ 20M	≤ 20M
Dead zone (event)	≤ 4M	≤ 4M	≤ 4M
dB readout resolution	0.01 dB		
Distance range (km)	0=32 km		
Data acquisition window (m or ft)	Selectable over full distance range		
Refractive index range	1.400 to 1.700		
Distance measurement accuracy	± 0.01% ± 0.5 m at 10 km		
	± 0.01% ± 5.0 m at 32 km		
Minimum length readout resolution	0.02 m or 1 in		
Horizontal scales	4 m (smallest window) to 32 km (largest window)		
Averaging time	Selectable (real time, 10 sec., 40 sec., continuous)		
Display	6 in diagonal screen, green		
Printer output	Integrated high speed graphics printer (optional)		
Connector types	Biconic, FC, D4, ST compatible, SMA, DINM		

Source: Courtesy of SIECOR, 1991.

Note: Specification tests were performed using 50-μm fiber with PC connectors. Specifications could vary with other fiber and connector types.

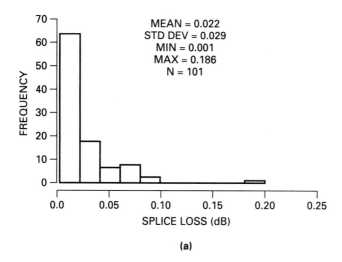

(a)

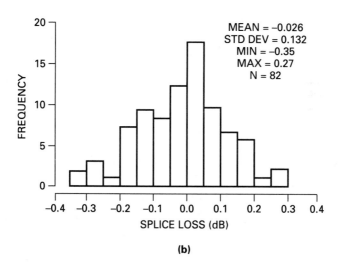

(b)

Figure 8.17   OTDR inaccuracy when measuring splice loss. (a) Histogram of splices that were made using the rotary splicing technique (AT&T) and measured using a direct measurement technique that provides accurate insertion loss, and (b) histogram shows the same splices measured with an OTDR. (Reprinted with permission of AT&T, all rights reserved.)

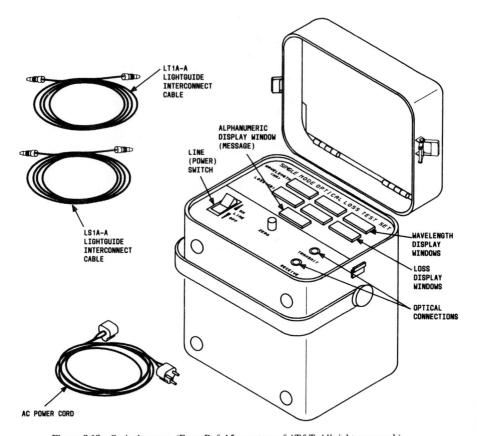

Figure 8.18   Optical test set. (From Ref. 15; courtesy of AT&T. All rights reserved.)

equipment. They are packaged in what appears like connector housings and simply connected in-line much like a jumper cable. Some fixed attenuators can be selectable or variable so that loss can be fine tuned for optimal performance.

Variable optical attenuators, on the other hand, have a calibrated optical loss variation mechanism and a means for connecting in-line. Some have a partially silvered mirror wheel that is rotated to dial up the required attenuation. Others have removable wafers that represent attenuation steps.

8.7.3 Cable-Plant Testing Process

8.7.3.1 Cable Inspection

Cable inspection checks the cables to ensure they are in proper condition for installation. It consists of testing the fiber for breaks and proper attenuation and sometimes bandwidth characteristics. Bandwidth characteristics are generally only

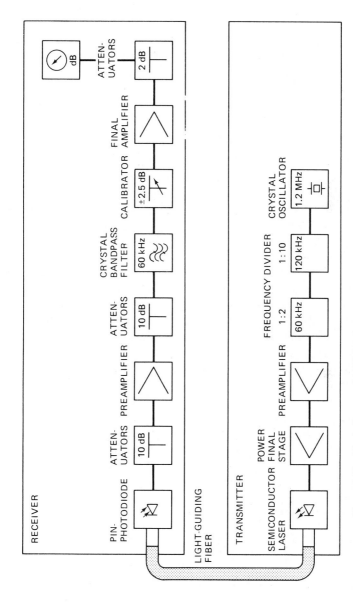

Figure 8.19 Block diagram of AEG-Telefunken attenuation test set. (Courtesy of AEG-Telefunken.)

checked on a sampling basis upon receiving the cable, and only for multimode fiber. With singlemode fiber, bandwidth is not a key inspection requirement since it is dictated more by source spectral width than by fiber manufacturing tolerances. There are generally two inspection stages involved.

Receiving Inspection.   In receiving inspection the objective is to determine that the manufacturer shipped what was ordered and that it meets the required specifications. It consists of visual inspection of cable and paperwork as well as some sampling tests of the fiber performance. Sampling tests check for attenuation of fibers and the bandwidth if these are deemed critical parameters. Precision test sets are used to perform these tests and the measurements are compared with shipping data. Such tests would normally be performed on 10% of the shipment, but could vary depending on circumstances.

Preinstallation Tests.   Preinstallation tests are performed before the reels are sent to the site for installation. They consist of (1) visual inspection of the reel to ensure flanges are not bent to the degree that they would interfere with installation; and (2) OTDR test of all fibers in all cables to check for broken fiber, attenuation anomalies mid span, cable length as specified, and attenuation within the expected range.

To perform these tests one end at least of every cable is prepared by stripping the cable jacket and exposing the buffered fibers. Each fiber is then cleaved with a tool to provide a good optical end finish. The fiber is then cleaned and inserted in the OTDR fiber holder and tested. The OTDR should be set for the same wavelength that the cable was factory tested at so that measurements can be compared. Fiber breaks will show up as fiber lengths shorter than the others. Attenuation bumps in the OTDR curve, midspan, could simply relate to factory fiber splices if the anomaly is very small. Large anomalies or those that follow some distance can be of concern. They may relate to a fiber defect or the cable pinching fibers. If they cause the fiber to be out of specification or if the anomaly is large, the reel should be rejected. Small anomalies should be rechecked when the fiber is loose or installed. They may have been due to microbending of the fiber on the tightly reeled cable.

8.7.3.2 Construction Testing

Construction testing consists of testing the quality of the splices as they are made along a span of cable. There are generally three quality tests: visual inspection, strength, and attenuation.

Visual Inspection.   After the splice is formed, particularly with a fusion splicer, the finished splice is inspected through the microscope before being removed or coated. A good splice is generally invisible or at a minimum a very thin line. If bubbles, a thick line of demarcation, or bulges appear at the splice point, the splice should be remade.

Strength.   A strength test is optional and the method somewhat subjective, but it does provide an added measure of insurance to the quality of the splice. Once the splice is performed, it is removed from the machine and given a slight tensile tug while it is held with the fingers. If cracks are present or the splice is weak it should break.

Attenuation.   Attenuation tests are performed in two ways, either the splicing machine performs a measurement using a local injection/detection relative measurement technique, or alternatively an optical time domain reflectometer.

With the local injection/detection technique, the splicing machine measures the actual or approximate attenuation of the splice by coupling light into one fiber and measuring light exiting the second spliced fiber, close to the splice point [23]. With this system, light is injected into the "transmit" fiber either end-on or through the cladding and buffer, and light is detected at the "receive" fiber either end-on or through the cladding and buffer (coupled by bending the fiber). After a good cleave is made of the fiber ends, the splicer automatically aligns the fiber cores with a matching fluid in between to get a baseline zero-loss measurement. After the fusion splice is made, a second relative measurement is made of optical power exiting the receive fiber and the difference is displayed as splice loss.

With the OTDR, the loss is measured after the splice is completed and removed from the machine (in a loss state). The OTDR is generally placed at one end of the cable span at a repeater or termination point, and coupled to the fibers by terminating that end of the cable with connectorized pigtails. Splicing then proceeds from that end on down the route to the next repeater point. As each splice is performed an OTDR signature of that splice can be made the instant that the splice is completed and the results can then be reported and permanently recorded. As Fig. 8.17 indicated, the OTDR results are very approximate on loss and therefore are only useful to determine whether the splice is of relatively good or bad quality. Using the OTDR in real time, however, gives instant feedback to the splice operator as to whether a splice has a signature above a certain loss threshold and therefore needs to be redone.

With the OTDR, owing to the nature of the way it measures loss with backscatter, to get a more accurate picture, loss must be measured in both directions. For this reason the OTDR must be moved to the opposite end of the cable span and the splice signatures redone in the opposite direction, before the splices are completely stowed and the enclosures sealed and secured.

Because of the nature of splice loss mechanisms in fiber, splice loss specifications should always be characterized statistically. Requiring that splicing crews meet an absolute maximum, particularly when measured with an OTDR, is not possible and simply invites trouble. As stated in Chapter 5, splice loss is a function of fiber tolerance mismatch and mechanical alignment mismatch. Adding to this the OTDR error, it is clear that we have three independent variables that interact in a way that can only be described with statistical probability. Only one of the three, alignment, is controllable to a degree by the splicing crews.

The best way to specify splice loss is to provide an average or mean loss that is to be obtained and then a standard deviation about that loss. The standard deviation describes the worst case loss you expect or desire for some percent of the fibers. One standard deviation describes the loss you expect to achieve in 84% of the cases for example. Three standard deviations is generally considered a worst case condition and represents 99.87% of the cases. By specifying the mean and standard deviation you create a picture of the expected distribution of loss results from splice to splice. The distribution follows what is known as a Normal curve. Such a distribution is illustrated in Fig. 7.6.

It is also valid to specify a maximum, usually the 3-standard deviation value, but this maximum should only be used as a threshold for determining whether a splice should be reattempted or not. A good practice is that if an individual splice exceeds a certain maximum it should be broken and respliced. If it again is greater than the maximum it should be reattempted. If on the third try it still exceeds the maximum, the condition of the splice equipment, cleaver, cleanliness of the environment, or operator situation should be checked. If all are satisfactory, then the splice should be left alone and the loss condition waived. At this point the excess loss may be due to fiber tolerance that cannot be improved by resplicing.

An example of a practical splice loss specification, based on OTDR measurement, is between 0.1 and 0.15 dB mean loss with an 0.1 SD and an 0.3 dB to 0.4 dB maximum for remake purposes. The lower figures would be practical only with active alignment splicing equipment.

8.7.3.3 Acceptance Testing the Optical Plant

Acceptance testing generally consists of an end-to-end attenuation test of the spliced and often connector terminated cable. Testing with connector terminations is generally the easiest and most accurate, although bare fiber tests are possible by using a mechanical or fusion splice to terminate the fiber ends to the attenuation test set. The objective of this test is to ensure that the installed cable span meets its specified transmission loss. It is not important to the system that each splice or length of fiber meet a specific loss as long as the end-to-end loss is acceptable.

A typical means of performing the acceptance test is to measure loss from one repeater location to another (or terminal end) using two optical test sets. One test set is used in the transmit configuration and the other in the receive. The configurations are then switched to obtain a second measurement in the reverse direction on the same fiber. The optical loss, based on the average of the two measurements minus a calibrated fixed loss for the test lead connectors, must be less than a specified limit to be accepted.

8.7.4 Optical Test of Electronics

Terminal equipment is accepted based on a set of parameters and functions peculiar to each set of equipment. These tests are specified either by the test procedure supplied by the manufacturer or that required by the system specification. Optical-power margin testing, however, is rather common to all equipment. It consists of a test of the power levels of the transmitter and a test of the optical power margin of the span between each transmitter and receiver.

Power levels are tested at the output of the transmitter optical connector, with the transmitter usually powered by a 50% duty cycle test signal, using a calibrated optical power meter.

Optical-power margin is measured by connecting a variable optical attenuator in the span between the transmitter and detector, generally on the detector end to prevent back-scatter into the laser. The equipment performance is measured using a BER tester with the equipment in the loop back mode. The attenuation is increased until the BER increases to the specification limit. The attenuation inserted is equivalent to the optical-power margin.

REFERENCES

1. E. Lacy, *Handbook of Electronic Safety Procedures* (Englewood Cliffs, N.J.: Prentice-Hall, Inc., 1977).

2. Amphenol RF Division Specification 349-502-3, Nov. 5, 1976.

3. Du Pont Safety Instruction, n.d.

4. "HFBR-0100, Fiber Optic Connector Assembly Tooling Kit," Hewlett-Packard User's Manual, Nov. 1980.

5. C. C. Timmermann, "Handling Optical Cables: Safety Aspects," *Applied Optics,* Vol. 16, No. 9, Sept. 1977, pp. 2380–2382.

6. CATV Cable Construction Mini-Manual, M/A-COM Comm/Scope Marketing Inc., P.O. Box 1729, Hickory, NC 28603.

7. CAPSCAN Trunk Cable Installation Tips, commercial brochures for CATV cable installation from CAPSCAN Corp.

8. American National Standard National Electrical Safety Code, ANSI C2, Institute of Electrical and Electronics Engineers, Inc., 345 E. 47th Street, New York, NY 10017.

9. D. Gibson, D. Pope, K. Wells, "Guidelines for Burying FO Cables," TELEPHONY, Jan. 30, 1984, Telephony Publishing Corp., 55 E. Jackson Blvd., Chicago, IL 60604.

10. "Specifications and Drawings for Underground Cable Installation," REA FORM 515d, 9/79, U.S. Department of Agriculture Rural Electrification Administration.

11. "Lightguide Premises Distribution System," AT&T Product Application Bulletin PAB-202A, Jan. 1985.

12. AT&T Technical Reference for Network Equipment Building Systems (NEBS), PUB-51001.

13. AT&T Product Application Bulletin PAB 202A, "Lightguide Premises Distribution System," Jan. 1985.

14. AT&T Practice 801-006-151, instruction on building fire stops.

15. AT&T Applications Bulletin 626-108-110, "Single-Mode Lightguide Cable," May 1986.

16. AT&T product literature, "Lightguide Universal Fiber Optic Closure," F-83AK8530, 8531.

17. Northern Telecom product literature, "NT6F26 Fiber Patch Panel Assembly."

18. "Optical Communications Products," RCA Publication OPT-115, June 1979.

19. J. Adams, C. Swinn, and G. Kavanagh, "Field Testing Fiber Optics," *Telephony,* May 14, 1979, pp. 29–31.

20. Paul Wendland, "Lighten the Burden of Fiber-Optic Measurements with New Instruments, Standards," *Electronic Design,* Vol. 27, No. 21, Oct. 11, 1979, pp. 126–130.

21. Michael K. Barnoski and S. D. Personick, "Measurements in Fiber Optics," *Proceedings of the IEEE,* Vol. 66, No. 4, April 1978, pp. 429–440.

22. M. Reynolds and P. Gagen, AT&T Bell Laboratories, "Field Splicing of Single-Mode Lightguide Cable," in *Science Systems and Devices for Communications,* (New York: IEEE/Elsevier Science Publishers B. V. [North Holland], 1984).

23. Furikawa Optical Fibers Systems "FITEL S-161" Fusion Splicer, technical brochure JE-015, The Furukawa Electric Co., Ltd.

Glossary*

Acceptance Angle: The angle measured from the longitudinal centerline up to the maximum acceptance angle of an incident ray that will be accepted for transmission along a fiber. The maximum acceptance angle is dependent on the indices of refraction of the two media that determine the critical angle. For a cladded glass fiber in air, the sine of the maximum acceptance angle is given by the square root of the difference of the squares of the indices of refraction of the fiber core glass and the cladding. *See* Maximum acceptance angle.

Acceptance Cone: A cone whose included apex angle is equal to twice the acceptance angle.

Analog-Intensity Modulation: In an optical modulator, the variation of the intensity (i.e., instantaneous output power level) of a light source in accordance with an intelligence-bearing signal or continuous wave, the resulting envelope normally being detectable at the other end of a [lightwave] transmission system.

Aperture: In an optical system, an opening or hole, through which light or matter may pass that is equal to the diameter of the largest entering beam of light that can travel completely through the system and that may or may not be equal to the aperture of the objective. *See* Numerical aperture.

*Adapted from "Vocabulary for Fiber Optics and Lightwave Communications," National Communication System (NCS), Technical Information Bulletin 79-1, Feb. 1979, prepared by the NCS Office of Technology and Standards. Deletions are indicated by ellipses. Changes and additions are indicated in brackets. References to "frequencies" have been changed to "wavelengths," and references to "microns" have been changed to "nanometers" and the units changed accordingly.

Avalanche Photodiode (APD):   A photo-detecting diode that is sensitive to incident photo energy by increasing its conductivity by exponentially increasing the number of electrons in its conduction-band energy levels through the absorption of the photons of energy, electron interaction, and an applied bias voltage. The photodiode is designed to take advantage of avalanche multiplication of photocurrent. As the reverse-bias voltage approaches the breakdown voltage, hole-electron pairs created by absorbed photons acquire sufficient energy to create additional hole-electron pairs when they collide with substrate atoms. Thus, a multiplication effect is achieved.

Blackbody: An ideal body that would absorb all radiation incident on it. When heated by external means, the spectral energy distribution of radiated energy would follow curves shown on optical spectrum charts. The ideal blackbody is a perfectly absorbing body. It reflects none of the energy that may be incident upon it. It radiates (perfectly) at a rate expressed by the Stefan-Boltzmann law and the spectral distribution of radiation is expressed by Planck's radiation formula. When in thermal equilibrium, an ideal blackbody absorbs perfectly and radiates perfectly at the same rate. The radiation will be just equal to absorption if thermal equilibrium is to be maintained. *Synonym:* ideal blackbody.

Brightness: An attribute or visual perception in accordance with which a source appears to emit more or less light. Since the eye is not equally sensitive to all colors, brightness cannot be a quantitative term. It is used in nonquantitative statements with reference to sensations and perceptions of light. . . .

Bundle: A group of . . . optical fibers . . . associated together and usually in a single sheath. Official fiber bundles are usually considered to be in a random arrangement and are used or considered as a single transmission medium. *See.* . . . Coherent bundle; Optical fiber bundle. . . .

Cable: A jacketed bundle or jacketed fiber, in a form that can be terminated. *See* . . . Fiber-optic cable. . . .

Cable Assembly: A cable terminated and ready for installation. . . .

Cable Core: The portion of a cable inside a common covering.

Cable Jacket: The outer protective covering applied over the internal cable elements.

Candela: The luminous intensity of 1/600,000 of a square meter of a blackbody radiator at the temperature of solidification of platinum, 2045 Kelvin. [A] 1-candela [point source] emits 4π lumens of light flux.

Candlepower: A unit of measure of the illuminating power of any light source, equal to the number of [candelas] of the source of light. A flux density of 1 lumen of luminous flux per steradian of solid angle measured from the source is produced by a point source of 1 candela emitting equally in all directions.

Chromatic Dispersion: Dispersion or distortion of a pulse in an optical waveguide due to differences in wave velocity caused by variations in the indices of refraction for different portions of the guide. . . .

Cladding: An optical conductive material with a lower refractive index placed over or outside the core material of . . . an optical fiber, or a thin film on a substrate, that serves to reflect or refract light waves so as to confine them to the core, and serves to protect the core. . . .

Coherent Bundle: A bundle of optical fibers in which the spatial coordinates of each fiber are the same or bear the same spatial relationship to each other at the two ends of the bundle. *Synonym:* aligned bundle. Used for imaging applications, not communications.

Coherent Light: Light that has the property that at any point in time or space, particularly over an area in a plane perpendicular to the direction of propagation or over time at a particular point in space, all the parameters of the wave are predictable and are correlated. . . .

Converging Lens: A lens that adds convergence to an incident bundle of light rays. One surface of a converging lens may be [convex] and the other plane plano-convex. Both may be convex (double-convex, biconvex) or one surface may be convex and the other concave (converging meniscus). *Synonyms:* convergent lens; convex lens. . . .

Core: The central primary light-conducting region of . . . an optical fiber, the refractive index of which must be higher than that of the cladding in order for the light waves to be internally reflected or refracted. Most of the optical power is in the core. *See also* Cladding; Cable core; Fiber core.

Coupler: In optical transmission systems, a component used to interconnect three or more optical conductors. *See,* . . . Data bus coupler; Non-reflective . . . coupler; Reflective star coupler; T coupler. . . .

Critical Angle: In terms of indices of refraction, the critical angle is the angle of incidence from a denser medium at an interface between the denser and less dense medium, at which all of the light is refracted along the interface (i.e., the angle of refraction is 90°). When the critical angle is exceeded, the light is totally reflected back into the denser medium. The critical angle varies with the indices of refraction of the two media with the relationship $\sin A\ (C) = n(2)/n(1)$, where $n(2)$ is the index of refraction of the less dense medium, $n(1)$ is the index of refraction of the denser medium, and $A(C)$ is the critical angle, as above. In terms of total internal reflection in an optical fiber, the critical angle is the smallest angle made by a meridional ray in an optical fiber that can be totally reflected from the innermost interface and thus determines the maximum acceptance angle at which a meridional ray can be accepted for transmission along a fiber. *See also* Total internal reflection.

Crosstalk: In an optical transmission system, leakage of optical power from one optical conductor to another. The leakage may occur by frustrated total reflection from inadequate cladding thickness or low absorptive quality. *See* Fiber crosstalk.

Cylindrical Lens: A lens with a cylindrical surface. . . .

Dark Current: The current that flows in a photodetector when there is no radiant energy or light flux incident upon its sensitive surface (i.e., total darkness). Dark current generally increases with increased temperature for most photodetectors.

Data Bus: In an optical communication system, an optical waveguide used as a common trunk line to which a number of terminals can be interconnected using optical couplers.

Data Bus Coupler: In an optical communication system, a component that interconnects a number of optical waveguides and provides an inherently bidirectional system by mixing and splitting all signals within the component.

Diffraction: The process by means of which the propagation of radiant waves or light waves are modified as the wave interacts with an object or obstacles. Some of the rays are deviated from their path by diffraction at the objects, whereas other rays remain undeviated by diffraction at the objects. As the objects become small in comparison with the wavelength, the concepts of reflection and refraction become useless and diffraction plays the dominant role in determining the redistribution of the rays following incidence upon the objects. Diffraction results in a deviation of light from the paths and foci prescribed by the rectilinear propagation prescribed by geometrical optics. Thus, even with a very small, distant source, some light, in the form of bright and dark bands, is found within a geometrical shadow because of the diffraction of the light at the edge of the object forming the shadow. Gratings with spacings of the order of the wavelength of the incident light cause diffraction. Such gratings can be ruled grids, spaced spots, or crystal lattice structures.

Diffraction Grating: An array of fine, parallel, equally spaced reflecting or transmitting lines that mutually enhance the effects of diffraction at the edges of each so as to concentrate the diffracted light very close to a few directions characteristic of the spacing of the lines and the wavelength of the diffracted light. If I is the angle of incidence, D the angle of diffraction, S the center-to-center distance between successive rulings, N the order of the spectrum, the wavelength is $L = (S/N) (\sin I + \sin D)$. If there is a large number of narrow, close, equally spaced rulings upon a transparent or reflecting substrate, the grating will be cable of dispersing incident light into its frequency component spectrum.

Diffusion: The scattering of light by reflection or transmission. Diffuse reflection results when light strikes an irregular surface such as a frosted window or the surface of a frosted or coated light bulb. When light is diffused, no definite image is formed.

Dispersion: (1) The process by which rays of light of different wavelength are deviated angularly by different amounts, as, for example, with prisms and diffraction gratings. (2) Phenomena that cause the index of refraction and other optical properties of a medium to vary with wavelength. Dispersion also refers to the frequency dependence of any of several parameters, for example, in the process by which an electromagnetic signal is distorted because the various frequency components of that signal have different propagation characteristics and paths. Thus, the components of a complex radiation are dispersed or separated on the basis of some characteristic. A prism disperses the components of white light by deviating each wavelength a different amount. *See* Chromatic dispersion; Fiber dispersion; Material dispersion; . . . Pulse dispersion. . . .

Diverging Lens: A lens that causes parallel light rays to spread out. One surface of a diverging lens may be concavely spherical and the other plane (planoconcave). Both may be concave (double concave) or one surface may be concave and the other convex (concave-convex, divergent-meniscus). The diverging lens is always thicker at the edge than at the center. *Synonyms:* concave lens; dispersive lens; divergent lens; negative lens. The diverging lens is considered to have a negative focal length.

Dopant: A material mixed, fused, amalgamated, crystallized, or otherwise added to another (intrinsic) material in order to achieve desired characteristics of the resulting material. For example, the germanium tetrachloride or titanium tetrachloride used to increase the refractive index of glass for use as an optical-fiber core material, or the gallium or arsenic added to silicon or germanium to produce a doped semiconductor for achieving donor or acceptor, positive or negative material for diode and transistor action.

Double Heterojunction: In a laser diode, two heterojunctions in close proximity, resulting in full carrier and radiation confinement and improved control of recombinations. [*See also* Heterojunction]

Double Heterojunction Diode: A laser diode that has two different heterojunctions, the difference being primarily in the stepped changes in refractive indices of the material in the vicinity of the *p-n* junction. The double heterojunction laser diode is widely used for [optical communication]. . . .

Edge-emitting LED: A light-emitting diode with a spectral output that emanates from between the heterogeneous layers (i.e., from an edge), having a higher output intensity and greater coupling efficiency to an optical fiber or integrated optical circuit than the surface-emitting LED, but not as great as the injection laser. Surface-emitting and edge-emitting LEDs provide several milliwatts of power in the spectral range 800–1200 nm at drive currents of 100 to 200 milliamperes; diode lasers at these currents provide tens of milliwatts. *See also* Surface-emitting LED.

Electromagnetic Spectrum: The entire range of wavelengths, extending from the shortest to the longest or conversely, that can be generated physically. This range of electromagnetic wavelengths extends almost from zero to infinity and includes the visible portion of the spectrum known as light. . . .

Electrooptic Effect: The change in the index of refraction of a material when subjected to an electric field. The effect can be used to modulate a light beam in a material, since many properties, such as light conducting velocities, reflection and transmission coefficients at interfaces, acceptance angles, critical angles, and transmission modes, are dependent upon the refractive indices of the media in which the light travels.

End-Fire Coupling: Optical-fiber and integrated optical-circuit (IOC) coupling between two waveguides in which the two waveguides to be coupled are butted up against each other. [It is] a more straightforward, simpler, and more efficient coupling method than evanescent field coupling. Mode pattern matching is required and accomplished by maintaining a unity cross-sectional area aspect ratio, axial alignment, and minimal lateral axial displacement. *See also* Evanescent-field coupling.

Evanescent-Field Coupling: Optical-fiber or integrated optical-circuit (IOC) coupling between two waveguides in which the two waveguides to be coupled are held parallel to each other in the coupling region. The evanescent waves on the outside of the waveguide enter the coupled waveguide bringing some of the light energy with it into the coupled waveguide. Close-to-core proximity or fusion is required. . . .

Exitance: *See* Radiant exitance.

Exit Angle: When a light ray emerges from a surface, the angle between the ray and a normal to the surface at the point of emergence. For an optical fiber, the angle between the output ray and the axis of the fiber. . . .

External Optical Modulation: Modulation of a [lightwave] in a medium by application of fields, forces, waves, or other energy forms upon a medium conducting a light beam in such a manner that a characteristic of either the medium, or the beam, or both are modulated in some fashion. External optical modulation can make use of such effects as the electrooptic, acoustooptic, magnetooptic, or absorption effect.

Fiber: *See* Graded-index fiber; Optical fiber; . . . Single-mode fiber; Step-index fiber.

Fiber Absorption: In an optical fiber, the [lightwave] power attenuation due to absorption in the fiber core material, a loss usually evaluated by measuring the power emerging at the end of successively shortened known lengths of the fiber.

Fiber Buffer: The material surrounding and immediately adjacent to an optical fiber that provides mechanical isolation and protection. Buffers are generally softer materials than jackets.

Fiber Cladding: A light-conducting material that surrounds the core of an optical fiber and that has a lower refractive index than the core material.

Fiber Core: The central light-conducting portion of an optical fiber. The core has a higher refractive index than the cladding that surrounds it.

Fiber Core Diameter: In an optical fiber, the diameter of the higher refractive index medium that is the primary transmission medium for the fiber.

Fiber Crosstalk: In an optical fiber, exchange of lightwave energy between a core and the cladding, the cladding and the ambient surrounding, or between differently indexed layers. Fiber crosstalk is usually undesirable, since differences in path length and propagation time can result in dispersion, reducing transmission distances. Thus, attenuation is deliberately introduced into the cladding by making it lossy.

Fiber-Detector Coupling: In fiber-optic transmission systems, the transfer of optical signal power from an optical fiber to a detector for conversion to an electrical signal. Many optical-fiber detectors have an optical-fiber pigtail for connection by means of a splice or a connector to a transmission fiber.

Fiber Diameter: The diameter of an optical fiber, normally inclusive of the core, the cladding if step-indexed, and any adherent coating not normally removed when making a connection, such as by a butted or tangential connection.

Fiber Dispersion: The lengthening of the width of an electromagnetic-energy pulse as it travels along a fiber; caused by material dispersion due to the frequency dependence of the refractive index; [by] modal dispersion, [due to] different group velocities of the different modes, and [by] waveguide dispersion due to frequency dependence of the propagation constant of that mode.

Fiber-Optic Cable: Optical fibers incorporated into an assembly of materials that provides tensile strength, external protection, and handling properties comparable to those of equivalent-diameter [electrical] cables. Fiber-optic cables (light guides) are a direct replacement for conventional coaxial cables and wire pairs.

Fiber-Optic Communications (FOC): Communication systems and components in which optical fibers are used to carry signals from point to point.

Fiber-Optic Multiport Coupler: An optical unit, such as a scattering or diffusion solid "chamber" of optical material, that has at least one input and two outputs, or at least two inputs and one output, that can be used to couple various sources to various receivers. The ports are usually optical fibers. If there is only one input and one output port, it is simply a connector.

Fiber-Optic Rod Coupler: A graded-index cylindrically shaped section of optical fiber or rod with a length corresponding to the pitch of the undulations of light waves caused by the graded refractive index, the light beam being injected via fibers at an off-axis end point on the radius, with the undulations of the resulting wave varying periodically from one point to another along the rod and with half-reflection layers at the 1/4-pitch point of the undulations providing for coupling between input and output fibers.

Fiber Optics: The technology of guidance of optical power, including rays and waveguide modes of electromagnetic waves along conductors of electromagnetic waves in the visible and near-visible region of the frequency spectrum, specifically when the optical energy is guided to another location through thin transparent strands. [Technology includes] conveying light or images through a particular configuration of glass or plastic fibers. Incoherent optical fibers will transmit light, as a pipe will transmit water, but not an image. [Aligned bundles of] optical fibers can transmit [a mosaic] image through perfectly aligned, small . . . clad optical fibers. Specialty fiber optics combine coherent and incoherent aspects. . . .

Fiber-Optic Splice: A nonseparable junction joining one optical conductor to another.

Fiber-Optic Transmission System: A transmission system utilizing small-diameter transparent fibers through which light is transmitted. Information is transferred by modulating the transmitted light. These modulated signals are detected by light-sensitive devices (i.e., photodetectors). *See* Laser fiber-optic transmission system.

Fiber-Optic Waveguide: A relatively long thin strand of transparent substance, usually glass, capable of conducting an electromagnet wave of optical wavelength (visible or near-visible region of the frequency spectrum) with some ability to confine longitudinally directed, or near-longitudinally directed light waves to its interior by means of internal reflection. The fiber-optic waveguide may be homogeneous or radially inhomogeneous with step or graded changes in its index of refraction, the indices being lower at the outer regions, the core thus being of increased index of refraction.

Filter: In an optical system, a device with the desired characteristics of selective transmittance and optical homogeneity, used to modify the spectral composition of radiant light flux. A filter is usually of special glass, gelatin, or plastic optical parts with plane parallel surfaces that are placed in the path of light through the optical system of an instrument to selectively absorb certain wavelengths of light, reduce glare, or reduce light intensity. Colored, ultraviolet, neutral density, and polarizing filters are in common use. Filters may be separate elements or integral devices mounted so that they can be placed in or out of position in a system as desired. . . .

Footcandle: A unit of illumination equal to 1 lumen incident per square foot. It is the illuminance [on] a surface placed 1 foot from a light source having a luminous intensity of 1 . . . candela. [Instead of footcandle, the preferred unit is the "lux," which is 1 lumen per square meter. One foot-candle = 10.764 lux.]

Fusion Splicing: In optical transmission systems using solid media, the joining together of two media by butting the media together, forming an interface between them, and then removing then common surfaces so that there be no interface between them when a light wave is propagated from one medium to the other, thus, [ideally] no reflection or refraction can occur at the former interface.

Gain-Bandwidth Product: The product of the [low-frequency] gain of an active device and [the half-power] bandwidth. For an avalanche photodiode, the gain-bandwidth product is the gain times the frequency of measurement when the device is biased for maximum obtainable gain.

Geometric Optics: The optics of light rays that follow mathematically defined paths in passing through optical elements such as lenses and prisms and optical media that refract, reflect, or transmit electromagnetic radiation. The branch of science that treats light propagation in terms of rays, considered as straight or curved lines in homogeneous and nonhomogeneous media.

Graded-Index Fiber: An optical fiber with a variable refractive index that is a function of the radial distance from the fiber axis, the refractive index getting progressively lower away from the axis. This characteristic causes the light rays to be continually refocused by refraction into the core. As a result, there is a designed continuous change in refractive index between the core and cladding along a fiber diameter. . . .

Graded-Index Profile: The condition of having the refractive index of a material, such as an optical fiber, vary continuously from one value at the core to another at the outer surface.

Heterojunction: In a laser diode, a boundary surface at which a sudden transition occurs in material composition across the boundary, such as a change in the refractive index as well as a change from a positively doped (p) region to a negatively doped (n) region (i.e., a p-n junction) in a semiconductor, or a positively doped region with a rapid change in doping level (i.e., a high concentration gradient of dopant versus distance), and usually at which a change in geometric cross section occurs and across which a voltage or voltage barrier may exist. Heterojunctions provide a controlled degree and direction of radiation confinement, there usually being an equal refractive index step at each heterojunction. *See* Double heterojunction. . . .

Homojunction: In a laser diode, a single junction (i.e., a single region of shift in doping from positive to negative majority carrier regions, or vice versa) and a change in refractive index, at one boundary, hence one energy-level shift, one barrier, and one refractive index shift.

Illuminance: Luminous flux incident per unit area of a surface. Illuminance is expressed in lumens per square meter. *Synonym:* illumination; [luminous incidance]

Incidence Angle: In optics, the angle between the normal to a reflecting or refracting surface and the incident ray.

Incident Ray: A ray of light that falls upon, or strikes, the surface of any object, such as a lens, mirror, prism, this printed page, the things we see, or the human eye. It is said to be incident to the surface.

Index-Matching Materials: Light-conducting materials used in intimate contact to reduce optical power losses by using materials with refractive indices at interfaces that will reduce reflection, increase transmission, avoid scattering, and reduce dispersion.

Infrared Band: The band of electromagnetic wavelengths between the extreme of the visible part of the spectrum, about 750 nm and the shortest microwaves, about 1,000,000 nm. . . .

Injection Laser Diode: A diode operating as a laser producing a monochromatic light modulated by injection of carriers across a *p-n* junction of a semiconductor [having] narrower spatial and wavelength emission characteristics for longer-range higher-data-rate systems than the LEDs, which are more applicable to larger-diameter and larger-numerical aperture fibers for lower-information bandwidths.

Insertion Loss: In [lightwave] transmission systems, the power lost at the entrance to a waveguide, such as an optical fiber or an integrated optical circuit, due to any and all causes. . . .

Integrated Optical Circuit: A circuit, or group in interconnected circuits, consisting of miniature solid-state optical components, such as light-emitting diodes, optical filters, photodetectors (active and passive), and thin-film optical waveguides on semiconductor or dielectric substrates. *Synonym:* optical integrated circuit.

Integrated Optics: The interconnection of miniature optical components via optical waveguides on transparent dielectric substrates, using optical sources, modulators, detectors, filters, couplers, and other elements incorporated into circuits analogous to integrated electronic circuits for the execution of various communication, switching, and logic functions.

Interference: In [lightwave] transmission, the systematic reinforcement [and/or] attenuation of two or more light waves when they are superimposed. Interference is an additive process. (The term is applied also to the converse process in which a given wave is split into two or more waves by, for example, reflection and refraction at beam splitters.) The superposition must occur on a systematic basis between two or more waves in order that the electric and magnetic fields of the waves can be additive and produce noticeable effects such as interference patterns. For example, the planes of polarizations should nearly or actually coincide or the wavelengths should [be] nearly or actually . . . the same.

Irradiance: The [radiant flux] per unit area of incident light upon a surface. The radiant flux incident upon a unit area of surface. It can be measured as watts per square meter, as for any form of electromagnetic waves, or as lumens per square meter when visible light is incident upon a surface. The old unit [for luminous incidance] was footcandles. *Synonym:* radiant flux density. . . .

Lambert: A unit of luminance, equal to $10^4/\pi$ [candelas] per square meter. . . .

Laser Diode: A junction diode, consisting a positive and negative carrier regions with a *p-n* transition region (junction) that emits electromagnetic radiation (quanta of energy) at optical wavelengths when injected electrons under forward bias recombine with holes in the vicinity of the junction. In certain materials, such as gallium arsenide, there is a high probability of radiative recombination producing emitted light, rather than heat, at a wavelength suitable for optical waveguides. Some light is reflected by the polished ends and is trapped to stimulate more emission, which further excites, overcoming losses, to produce laser action. *See* Injection laser diode.

Laser Fiber-Optic Transmission System: A system consisting of one or more laser transmitters and associated fiber-optic cables. During normal operation, the laser radiation is limited to the cable. . . .

Laser Line Width: In the operation of a laser, the wavelength range over which most of the laser beam's energy is distributed.

Launch Angle: In an optical fiber . . . the angle between the input radiation vector (i.e., the input light chief ray) and the axis of the fiber. . . . If the ends of the fibers are perpendicular to the axis of the fibers, the launch angle is equal to the angle of incidence when the ray is external . . . and the angle of refraction when initially inside the fiber.

Lens: An optical component made of one or m ore pieces of a material transparent to the radiation passing through. Having curved surfaces, it is capable of forming an image, either real or virtual, of the object source of the radiation, at least one of the curved surfaces being convex or concave, normally spherical but sometimes aspheric. *See* . . . Converging lens; Diverging lens.

   A transparent optical element, usually made from optical glass, having two opposite polished major surfaces of which at least one is convex or concave in shape and usually spherical. The polished major surfaces are shaped to that they serve to change the amount of convergence or divergence of the transmitted rays. . . .

Light: . . . Radiant electromagnetic energy within the limits of human visibility and therefore with wavelengths to which the human retina is responsive. Approximately 380 to 780 nm. . . .

Light-Emitting Diode (LED):   A diode that [has applications] similar to [those of] a laser diode, with the same output power level, the same output limiting modulation rate, and the same operational current densities . . . but [has] greater simplicity, tolerance, and ruggedness; and about 10 times the spectral width of [laser diode] radiation.

Light Ray: A line, perpendicular to the wavefront of lightwaves, indicating their direction of travel and representing the lightwave itself.

[Lightwave] Communications: That aspect of communications and telecommunications devoted to the development and use of equipment that uses electromagnetic waves in or near the visible region of the spectrum for communication purposes. [Lightwave] communication equipment includes sources, modulators, transmission media, detectors, converters, integrated optic circuits, and related devices, used for generating and processing lightwaves. The term "optical communications" is oriented toward the notion of optical equipment, whereas the term "[lightwave] communications" is oriented toward the signal being processed. *Synonym:* optical communications. *See also* Light.

Lumen: The SI unit of [luminous] flux corresponding to $1/(4\pi)$ of the total [luminous] flux emitted by a source having an intensity of 1 candela. . . .

Luminance: The . . . luminous intensity [per unit area] emitted by a light source in a given direction by an infinitesimal area of the source. . . . Luminance is usually stated as luminous intensity per unit area (i.e., luminous flux emitted per unit solid angle projected per unit projected area. . . .

Luminous Flux: The quantity that specifies the capacity of the radiant flux . . . to produce . . . visual sensation known as brightness. Luminous flux is radiant flux [weighted by] its luminous [efficacy]. Unless otherwise stated, luminous flux pertains to the standard photooptic observer.

Luminous Intensity: The ratio of the luminous flux emitted by a light source, or an element of the source, in an infinitesimally small cone about the given direction, to the solid angle of that cone, usually stated as luminous flux emitted per unit solid angle.

Lux: A unit of illuminance equal to a lumen incident per square meter of surface. . . .

Magneto-optic Effect: The rotation of the plane of polarization of plane-polarized light waves in a medium brought about when subjecting the medium to a magnetic field (Faraday rotation). The effect can be used to modulate a light beam in a material since many properties, such as conducting velocities, reflection and transmission coefficients at interfaces, acceptance angles, critical angles, and transmission modes, are dependent upon the direction of propagation at interfaces in the media in which the light travels. . . . The magnetic field is in the direction of propagation of the light wave. *Synonym:* Faraday effect.

Material Dispersion: [Bandwidth limitation owing to variable material property in] an optical transmission media used in optical waveguides, such as the variation in the refractive index of a medium as a function of wavelength, optical fibers, slab dielectric waveguides, and integrated optical circuits. Material dispersion contributes to group-delay distortion, along with waveguide-delay distortion and multimode group-delay spread. The part of the total dispersion of an electromagnetic pulse in a waveguide caused by the changes in properties of the material with which the waveguide such as an optical fiber is made, due to changes in frequency. As wavelength increases . . . material dispersion decreases; at [short wavelengths] the rapid interactions of the electromagnetic field with the waveguide material (optical fiber) renders the refractive index even more dependent upon wavelength.

Maximum Acceptance Angle: The maximum angle between the longitudinal axis of an optical transmission medium, such as an optical fiber or a deposited optical film, and the normal to the wavefront [at which propagation can take place] (i.e., the direction of the entering light ray), in order that there be total internal reflection of the portion of incident light that is transmitted through the fiber interface (i.e., the [input angle for which] the angle between the transmitted ray and the normal to the inside surface of the cladding is greater than the critical angle). The [sine of the] maximum acceptance angle is given by the square root of the difference of the squares of the indices of refraction of the fiber core glass and the cladding. The square root of the difference of the square is called the numerical aperture (NA).

Microbending Loss: In an optical fiber, the loss or attenuation in signal power caused by small bends, kinks, or abrupt discontinuities in direction of the fibers, usually caused by fiber cabling or by wrapping fibers on drums. Microbending losses usually result from a coupling of guided modes among themselves and among the radiation modes.

Mirror: A flat surface optically ground and polished on a reflecting material, or a transparent material that is coated to make it reflecting, used for reflecting light. A beam-splitting mirror has a lightly deposited metallic coating that transmits a portion of the incident light and reflects the remainder. A smooth highly polished plane or curved surface for reflecting light. Usually, a thin coating of silver or aluminum on glass constitutes the actual reflecting surface. When this surface is applied to the front face of the glass, the mirror is a front-surface mirror. . . .

Monochromatic: Pertaining to a composition of one color. Purely monochromatic light has all its energy confined to one . . . wavelength.

Monochromatic Light: Electromagnetic radiation, in the visible or near-visible (light) portion of the spectrum, that has only one frequency or wavelength.

Monochromatic Radiation: Electromagnetic radiation that has one . . . wavelength. . . .

Multichannel Cable: In optical-fiber systems, two or more cables combined in a single jacket, harness, strength member, cover, or other unitizing element.

Near Infrared: Pertaining to electromagnetic wavelengths from 750 to 3000 nm.

Noise-equivalent Power (NEP): In optics, the [bandwidth-normalized (per square root of bandwidth) value of optical power required to produce unity rms signal-to-noise ratio. NEP is a common parameter in specifying detector performance [in watts per root hertz]. NEP is useful for comparison only if modulation frequency, bandwidth, detector, area, and temperature are specified. NEP is indicated in watts [per root hertz].

Nonreflective Star Coupler: An optical-fiber coupling device that enables signals in one or more fibers to be transmitted to one or more other fibers by entering the input signal fibers into an optical-fiber volume without an internal reflecting surface so that the diffused signals pass directly to the output fibers on the opposite side of the fiber volume for conduction away in one or more of the output fibers. The optical-fiber volume is a shaped piece of the optical-fiber material to achieve transmission of two or more inputs to two or more outputs. *See also* Reflective star coupler; T coupler.

Numerical Aperture (NA): A measure of the light-accepting property of an optical fiber (e.g., glass), given by NA = square root of the difference of the squares of the indices of refraction of the core n (1), and the cladding, $n(2)$. If $n(1)$ is 1.414 (glass) and $n(2)$ is 1.0 (air), the numerical aperture is 1.0, and all incident rays will be trapped. The numerical aperture is a measure of the characteristic of an optic conductor in terms of its acceptance of impinging light. The degree of openness, light-gathering ability, angular acceptance, and acceptance cone are all terms describing this characteristic. It may be necessary to specify that the indices of refraction are for step-index fibers and for graded-index fibers $n(1)$ is the maximum index in the core and $n(2)$ is the minimum index in the cladding. As a number, the NA expresses the light-gathering power of a fiber. It is mathematically equal to the sine of the acceptance angle. . . . The numerical aperture is also equal to the sine of the half-angle of the widest [conical] bundle of rays capable of entering a lens, multiplied by the index of refraction of the medium containing that bundle of rays (i.e., the incident medium).

Optical Attenuator: In an optical-fiber data link or integrated optical circuit, a device used to reduce the intensity (i.e., attenuate the [lightwaves] when inserted into an optical waveguide). Three basic forms of optical attenuators have been developed: a fixed optical attenuator, a stepwise variable optical attenuator, and a continuous variable optical attenuator. One form of attenuator uses a filter consisting of a metal film evaporated onto a sheet of glass to obtain the attenuation. The filter might be tilted to avoid reflection back into the input optical fiber or cable. . . .

Optical Directional Coupler: A device used in optical-fiber communication systems, such as CATV and data links for optical-fiber measurements, to combine or split optical signals at desired ratios by insertion into a transmission line, for example, a three-port or four-port unit with precise connectors at each port to enable inputs to be coupled together and transmitted via multiple outputs.

Optical Fiber: A single discrete optical transmission element usually consisting of a fiber core and a fiber cladding. As a light-guidance system (dielectric waveguide) that is usually cylindrical in shape, it consists either of a cylinder of transparent dielectric material of given refractive index whose walls are in contact with a second dielectric material of a lower refractive index, or of a cylinder whose core has a refractive index that gets progressively lower away from the center. The length of a fiber is usually much greater than its diameter. The fiber relies upon internal reflection to transmit light along its axial length. Light enters one end of the fiber and emerges from the opposite end with losses dependent upon length, absorption scattering, and other factors. . . .

Optical-Fiber Bundle: Many optical fibers in a single protective sheath or jacket. The jacket is usually polyvinyl chloride (PVC). The number of fibers might range from a few to several hundred, depending on the application and the characteristics of the fibers.

Optical-Fiber Coating: A protective material bonded to an optical fiber, over the cladding if any, to preserve fiber strength and inhibit cabling losses, by providing protection against mechanical damage, protection against moisture and debilitating environments, compatibility with fiber and cable manufacture, and compatibility with the jacketing process. Coatings include fluorpolymers, Teflon, Kynar, polyurethane, and many others. . . .

Optical Repeater: An optical/optical, optical/electrical, or electrical/optical signal amplification and processing device.

Optical Surface: In an optical system, a reflecting or refracting surface of an optical element, or any other identified geometric surface in the system. Normally, optical surfaces occur at surfaces of discontinuity (abrupt changes) of fractive indices, absorptive qualities, transmissivity, vitrification, or other optical quality or characteristic.

Optical Transmitter: A source of light capable of being modulated and coupled to a transmission medium such as an optical fiber or an integrated optical circuit.

Optics: That branch of physical science concerned with the nature and properties of electromagnetic radiation and with the phenomena of vision. . . .

Optoelectronic Device: (1) A device [that emits or is] responsive to electromagnetic radiation in the visible, infrared, or ultraviolet spectral regions of the frequency spectrum [converting electric signals to optical, or vice versa] . . . or utilizes such electromagnetic radiation for its internal operation. The wavelengths handled by these devices range from approximately 300 to 30,000 nm. (2) Electronic devices associated with light, serving as sources, conductors, or detectors. . . .

Photodetector: A device capable of [detecting or] extracting the information from an optical carrier, (i.e., a thermal detector or a photon detector, the latter being used for communications more than the former. . . .

Photodetector Responsivity: The ratio of the rms value of the output current or voltage of a photodetector to the rms value of the . . . optical power input. In most cases, detectors are linear in the sense that the responsivity is independent of the intensity of the incident radiation. Thus, the detector response in amps or volts is proportional to incident optical power, watts. . . .

Photon: A quantum of electromagnetic energy. The energy of a photon is $[h/c/\lambda]$, where h is Planck's constant, c is the speed of light, and λ is the wavelength]. . . .

PIN Diode: A junction diode doped in the forward direction positive, intrinsic, and negative, in that order. PIN diodes are used as photodetectors in fiber and integrated optical circuits.

Prism: A transparent body with at least two polished plane faces inclined with respect to each other, from which light is reflected or through which light is refracted. When light is refracted by a prism whose refractive index exceeds that of the surrounding medium, it is deviated or bent toward the thicker part of the prism. . . .

Pulse Dispersion: A separation or spreading of input optical signals along the length of a transmission line, such as an optical fiber. This limits the useful transmission bandwidth of the fiber. It is expressed in time and distance as nanoseconds per kilometer. Three basic mechanisms for dispersion are the material effect, the waveguide effect, and the multimode effect. Specific causes include surface roughness, presence of scattering centers, bends in the guiding structure, deformation of the guide, and inhomogeneities of the guiding medium. *Synonym:* pulse spreading.

Radiance: The radiant intensity of electromagnetic radiation per unit projected area of a source or other area (i.e., it is the radiant power of electromagnetic radiation per unit solid angle and per unit surface area normal to the direction considered). The surface may be that of a source detector, or it may be any other real or virtual surface intersecting the flux. The unit of measure is watts/steradian-square meter. . . . *Synonym:* emittance. . . .

Radiant Exitance: The radiant power emitted [in all directions] by a unit area of source.

Radiant Flux: The time rate of flow of radiant energy. The units are watts, or joules/second. The radiant energy crossing or striking a surface per unit time, usually measured in watts.

Radian Intensity: The radiant power per unit solid angle in the direction considered (i.e., the time rate of transfer of radiant energy per unit solid angle, or the flux radiated per unit solid angle about a specified direction). The unit of measure is watts/steradian or joules/(steradian-second). . . .

Radiant Power: The time rate of flow of electromagnetic energy. The unit is watts or joules/second. [The preferred term is "radiant flux."]

Radiant Transmittance: The ratio of the radiant flux transmitted by an object to the incident radiant flux.

Radiation: The electromagnetic waves or photons emitted from a source. . . .

Radiometry: The science devoted to the measurement of radiated electromagnetic [flux or flux per unit area]. In [lightwave] communications and the use of optical fibers, primary concern is devoted to radiometry rather than photometry. [Photometry deals with measurement of optical radiation in the visible spectrum (i.e., luminous flux).]

Reflectance: The ratio of the reflected flux to the incident flux. This term is applied to radiant and to luminous flux. Unless qualified, reflectance applies to specular, or regular, reflection. . . .

Reflection: When electromagnetic waves, more appropriately light rays, strike a smooth polished surface, their return or bending back into the medium from whence they came. Specular or regular reflection from a polished surface, such as a mirror, will return a major portion of the light in a definite direction lying in the plane of the incident ray and the normal after specular reflection. Light can be made to form a sharp image of the original source. Diffuse reflection occurs when the surface is rough and the reflected light is scattered from each point in the surface. These diffuse rays cannot be made to form an image of the original source, only of the diffusely reflecting surface itself. *See* Total internal reflection; Snell's law.

Reflection Angle: When a ray of electromagnetic radiation strikes a surface, and is reflected in whole or in part by the surface, the angle between the normal to the reflecting surface and the reflected ray. *See* Critical angle.

Reflective Star Coupler: An optical-fiber coupling device that enables signals in one or more fibers to be transmitted to one or more other fibers by [injecting] the signals into one side of an optical cylinder, fiber, or other piece of material, with a reflecting back surface to as to reflect the diffused signals back to the output ports on the same side of the material, for conduction away in one or more fibers. *See also* T coupler. . . .

Refraction: The bending of oblique (nonnormal) incident electromagnetic waves or rays as they pass from a medium of one index of refraction into a medium of a different index of refraction, coupled with the changing of the velocity of propagation of the electromagnetic waves when passing from one medium to another with different indices of refraction. The waves or rays are usually changed in direction (i.e., bent) crossing the media interface. *See* Refractive index; Snell's law.

Refraction Angle: When an electromagnetic wave strikes a surface and is wholly or partially transmitted into the new medium, of which the struck surface is the boundary, the acute angle between the normal to the refracting surface at the point of incidence, and the refracted ray.

Refractive Index: (1) The ratio of the velocity of light in a vacuum to the velocity of light in the medium whose index of refraction is desired, for example, n = [1.6] for certain kinds of glass. (2) [Relative refractive index is the] ratio of the sines of the angle of incidence and the angle of refraction when light passes from one medium to another. The index between two media is the relative index, while the index when the first medium is a vacuum is the absolute index of the second medium. The index of refraction expressed in tables is the absolute index, that is, vacuum to substance at a certain temperature, with light of a certain wavelength. Examples, vacuum, 1.000; air, 1.000292; water, 1.333; ordinary crown glass, 1.516. Since the index of air is very close to that of vacuum, the two are often used interchangeably. *Synonyms:* absolute refractive index; index of refraction.

Single-Mode Fiber: A fiber waveguide that supports the propagation of only one mode. The single-mode fiber is usually a low-loss optical waveguide with a very small core. . . . It [generally] requires a laser source for the input signals because of the very small entrance aperture (acceptance cone). The small core radius approaches the wavelength of the source; consequently, only a single mode is propagated. [Mode is, in simple terms, the path of an optic ray.]

Skew Ray: In an optical fiber, a light ray that never intersects the axis of the fiber while being internally reflected. The skew ray is at an angle to the fiber axis. If the fiber waveguide is straight, a skew ray traverses a helical path along the fiber, not crossing the fiber axis. A skew ray is not confined to the meridian plane. The skew ray is not a meridional ray.

Snell's Law: When electromagnetic waves such as light pass from a given medium to a denser medium, its path is deviated toward the normal; when passing into a less dense medium, their path is deviated away from the normal. Snell's law, often called the law of refraction, defines this phenomenon by describing the relation between the angle of incidence and the angle of refraction as $\sin i / \sin r = n(r)/n(i)$, where i is the angle of incidence, r is the angle of refraction, $n(r)$ is the refractive index of the medium containing the refracted ray, and $n(i)$ is the refractive index containing the incident ray. Stated in another way, both laws, that of reflection and of refraction, are attributed to Snell: namely, when the incident ray, the normal to the surface at the point of incidence of the ray on the surface, the reflected ray, and the refracted ray all lie in a single plane. The angle between the incident ray and the normal is equal in magnitude to the angle between the reflected ray and the normal. The ratio of the sine of the angle between the normal and the incident ray to the sine of the angle between the normal and the refracted ray is a constant. *See also* Refraction.

Solid-State Laser: A laser whose active medium is a solid material such as glass, crystal, or semiconductor material rather than gas or liquid.

Source-Coupler Loss: In an optical data link, optical communication system, or optical-fiber system, the loss, usually expressed in dB, between the light source and the device or material that couples the light source energy from the source to the fiber cable.

Source-Fiber Coupling: In fiber-optic transmission systems, the transfer of optical signal power emitted by a light source into an optical fiber, such coupling being dependent upon many factors, including geometry and fiber characteristics. Many optical-fiber sources have an optical-fiber pigtail for connection by means of a splice or a connector to a transmission fiber.

Source-to-Fiber Loss: In an optical fiber, signal power loss caused by the distance of separation between a signal source and the conducting fiber.

Spectral Bandwidth: The wavelength interval in which a radiated spectral quantity is a specified fraction of its maximum value. The fraction is usually taken as 0.50 of the maximum power level. . . . If the electromagnetic radiation is light, it is the radiant intensity half-power points that are used.

Spontaneous Emission: In a laser, the emission of light that does not bear an amplitude, phase, or time relationship with an applied signal and is therefore a random noise-like form of radiation.

Step-Index Fiber: A fiber in which there is an abrupt change in refractive index between the core and cladding along a fiber diameter, with the core refractive index higher than the cladding refractive index. These may be more than one layer, each layer with a different refractive index that is uniform throughout the layer, with decreasing indices in the outside layer.

Step-Index Profile: The condition of having the refractive index of a material, such as an optical fiber, change abruptly from one value to another at the core-cladding interface, or at other interfaces if several layers are present.

Steradian: The unit of solid angular measure; [for a solid angle with its apex at the center of a sphere, the measure in steradians is] the subtended surface area of [the] sphere divided by the square of the sphere radius. There are 4π steradians in a sphere. The solid angle subtended by a cone of half-angle A is $2\pi(1 - \cos A)$ steradians.

Surface-Emitting LED: A light-emitting diode with a spectral output that emanates from the surface of the layers, having a lower output intensity and lower coupling efficiency to an optical fiber or integrated optical circuit than the edge-emitting LED and the injection laser. Surface-emitting LEDs provide several milliwatts of power in the spectral range 800 to 1200 nm at drive currents of 100 to 200 milliamperes; diode lasers at these currents provide tens of milliwatts. *Synonyms:* front-emitting LED; Burrus LED. *See also* Edge-emitting LED.

T Coupler: In an optical fiber, a reflective surface placed inside the fiber, at 45 degrees to the direction of wave propagation, allowing a part of the signal power to be reflected from one side of the surface out of the fiber at right angles in one direction, and an input signal from the other side of the fiber to be reflected from the other side of the 45-degree reflective surface so as to propagate in the fiber, longitudinally, in the same direction as the original signal to which the input signal is being added and the output signal is being taken. Two tee-couplers can be combined in a single unit for input and output of signals in both directions of propagation. In addition to an optical component used to interconnect a number of terminals through optical waveguides by using partial reflections at dielectric interfaces or metallic surfaces, coupling can be accomplished simply by splitting the waveguide bundle so that fractions can diverge in different directions. See also Reflective star coupler, Nonreflective star coupler.

Thick Lens: A lens whose axial thickness is so larger that the principal points and the optical center cannot be considered as coinciding at a single point on its optical axis.

Thin-Film Optical Modulator: A device made of multilayered films of material of different optical characteristics capable of modulating transmitted light by using electro-optic, electroacoustic, or magneto-optic effects to obtain signal modulation. Thin-film optical modulators are used as component parts of integrated optical circuits.

Thin-Film Optical Multiplexer: A multiplexer consisting of layered optical materials that make use of electro-optic, electroacoustic, or magneto-optic effects to accomplish the multiplexing. Thin-film optical multiplexers may be component parts of integrated optical circuits.

Thin-Film Optical Switch: A switching device for performing logic operations using light waves in thin films, usually supporting only one propagation mode, making use of electro-optic, electroacoustic, or magneto-optic effects to perform switching functions, such as are performed by semiconductor gates (*AND, OR,* negation). Thin-film optical switches may be component parts of integrated optical circuits.

Thin-Film Optical Waveguide: An optical waveguide consisting of thin layers of differing refractive indices, the lower indexed material on the outside or as a substrate, usually for supporting a single electromagnetic wave propagation mode with laser sources. The thin-film waveguide lasers, modulators, switches, directional couplers, filters, and related components need to be coupled from their integrated optical circuits to the optical waveguide transmission media, such as optical fibers and slab dielectric waveguides.

Thin Lens: A lens whose axial thickness is sufficiently small that the principal points, the optical center, and the vertices of its two surfaces can be considered as coinciding at the same point on its optical axis.

Total Internal Reflection: The reflection that occurs within a substance because the angle of incidence of light striking the boundary surface is in excess of the critical angle. *See also* Critical angle.

Transmittance: The ratio of the flux that is transmitted through an object, to the incident radiant or luminous flux. Unless qualified, the term is applied to regular (i.e., specular) transmission. . . .

Waveguide Delay Distortion: In an optical waveguide, such as an optical fiber, dielectric slab waveguide, or an integrated optical circuit, the distortion in received signal caused by the differences in propagation time for each wavelength (i.e., the delay versus wavelength effect for each propagating mode), causing a spreading of the total received signal at the detector. Waveguide delay distortion contributes to group-delay distortion, along with material dispersion and multimode group-delay spread.

Waveguide Dispersion: The part of the total dispersion attributable to the dimensions of the waveguide since they are critical for modes allowed and not allowed to propagate, such that waveguide dispersion increases as frequency decreases, due to these dimensions and their variation along the length of the guide.

Wavelength: The length of a wave measured from any point on one wave to the corresponding point on the next wave, such as from crest to crest. Wavelength determines the nature of the various forms of radiant energy that comprise the electromagnetic spectrum; for example, it determines the color of light. . . . [Wavelengths of particular sources are usually given as their wavelengths in free space.]

Wavelength Division Multiplex (WDM):   In optical communication systems, the multiplexing of light waves in a single medium such as a bundle of fibers, such that each of the waves is of a different wavelength and is modulated separately before insertion into the medium. Usually, several sources are used, such as a laser, or a dispersed white source, each having a distinct center wavelength. WDM is the same as frequency-division multiplexing (FDM) applied to other than visible light frequencies of the electromagnetic spectrum.

INDEX